Independent Energy Regulation in a Developing Economy

Sudha Mahalingam · Kapilan Mahalingam

Independent Energy Regulation in a Developing Economy

Stakeholder Perspectives and Legal Interpretations

Springer

Sudha Mahalingam
National Institute of Advanced Studies
Bengaluru, Karnataka, India

Kapilan Mahalingam
Graduate School
Pennsylvania State University
University Park, PA, USA

ISBN 978-981-99-4397-5 ISBN 978-981-99-4398-2 (eBook)
https://doi.org/10.1007/978-981-99-4398-2

This Springer imprint is published by the registered company Springer Nature Singapore Pte Ltd.
The registered company address is: 152 Beach Road, #21-01/04 Gateway East, Singapore 189721, Singapore

To Mali and Girish, The Silent Sentinels

Acknowledgements

Sometimes, having someone to look over your shoulder just in case you veered off course can be a blessing. While I was figuring out how to organise and present my ideas, Kapilan took on this role so persistently and persuasively that I made him complicit in the enterprise. I am indeed grateful to him for agreeing to share the burden and the blame, if there are any. Others came along and shared their perspectives from time to time, and I am grateful to all of them. Professor R. Srikanth and Prof. A. V. Krishnan, my colleagues at NIAS with whom I brainstormed ideas, Daljit Singh who patiently went through my draft chapters and pointed out the non-sequiturs and slips, but also shared his valuable insights on the difficulties of regulating a government-owned utility, Shantanu Dixit at Prayas who ploughed through the legalese in the chapters and reined me in from making generalizations and Prasad V. Jayaram who graciously added an outsider's perspective to my chapters. Finally, Dr. Usha Ramachandra's insights added value to our analyses and we are indeed grateful to her. I am also grateful to NIAS doctoral scholar Soumyadeep Das for collating data that has buttressed some of our arguments in the book. This work would not have been possible but for the space and freedom so generously allowed me by my employer, National Institute of Advanced Studies. And my gratitude to uncomplaining husband Mali who put up with all the chaos it caused at home as I buried myself into the task, is immense. All errors remain mine, of course.

Sudha Mahalingam

Contents

About the Authors

Sudha Mahalingam's association with energy sector goes back three decades and straddles the entire spectrum of professional engagements—as regulator, academic, advisor, consultant, trainer of regulators and business journalist reporting on India's rapidly transforming energy sector in the first decade of economic liberalization. She is currently Adjunct Professor at the National Institute of Advanced Studies in Bangalore where during her three year tenure as Dr. Raja Ramanna Chair Professor in 2022. She taught regulatory economics to graduate students. She has been full-time member of India's Petroleum and Natural Gas Regulatory Board, energy member of India's National Security Advisory Board, Visiting Fellow at the Elliott School of International Affairs, George Washington University, USA, member of the Scientific Advisory Board of Delft University in The Netherlands and a senior fellow at Nehru Memorial Museum and Library, New Delhi. Sudha has published in peer-reviewed and scholarly journals, contributed chapters to books on energy, represented India in international conferences and lectured at prestigious institutions. Sudha holds a doctorate in regulatory economics from University of Technology, Sydney, an M.Phil. in applied economics from Jawaharlal Nehru University, New Delhi and a Bachelor of Law from Delhi University.

Kapilan Mahalingam is a doctoral student of economics at Penn State University, USA. He is currently working on Industrial Organization which includes Regulatory Economics and focuses on identifying optimal incentives for renewable energy in India. He also works on Information Theory and Big Data. His earlier work examined shared incentives for crowdfunding, land acquisition and pollution estimation. Kapilan has a Masters in Economics from Delhi School of Economics and undergraduate degree in economics from Delhi University.

Abbreviations

APCPDCL	Andhra Pradesh Central Power Distribution Corporation Limited
APERC	Andhra Pradesh Electricity Regulatory Commission
APL	Adani Power Limited
APM	Administered Pricing Mechanism
APMuL	Adani Power Mundra Limited
APSEB	Andhra Pradesh State Electricity Board
APSPDCL	Andhra Pradesh Southern Power Distribution Corporation Limited
APTEL	Appellate Tribunal for Electricity
BEST	Brihanmumbai Electric Supply and Transport
BPCL	Bharat Petroleum Corporation Limited
BSES	Bombay Suburban Electric Supply
CEA	Central Electricity Authority
CERC	Central Electricity Regulatory Commission
CGD	City Gas Distribution
CGPL	Coastal Gujarat Power Limited
CNG	Compressed Natural Gas
CSA	Coal Supply Agreement
DPR	Detailed Project Report
ECF	Energy Charges Formula
ERC	Electricity Regulatory Commission
FERC	Federal Energy Regulatory Commission
FPC	Federal Power Corporation
GAIL	Gas Authority of India Limited
GCV	Gross Calorific Value
GERC	Gujarat Electricity Regulatory Commission
GMDC	Gujarat Mineral Development Corporation
GoI	Government of India
GUVNL	Gujarat Urja Vidyut Nigam Limited
HNPCL	Hinduja National Power Corporation Limited
HPC	High Powered Committee
HPCL	Hindustan Petroleum Corporation Limited

ICC	Interstate Communications Commission
IGL	Indraprastha Gas Limited
IMF	International Monetary Fund
IPP	Independent Power Producer
IRA	Independent Regulatory Agencies
LoA	Letter of Agreement
MIAL	Mumbai International Airport Limited
MMTPA	Million Metric Tonnes Per Annum
MoEF	Ministry of Environment and Forests
MoP	Ministry of Power
MoU	Memorandum of Understanding
MRTPC	Monopolies and Restrictive Trade Practices Commission
MSDCL	Maharashtra State Electricity Distribution Co. Ltd.
MSEB	Maharashtra State Electricity Board
NCLT	National Company Law Tribunal
NGO	Non-Government Organization
NHPC	National Hydropower Corporation
NPC	Nuclear Power Corporation
NPL	Nabha Power Limited
NTPC	National Thermal Power Corporation
O&M	Operation and Maintenance
OFGEM	Office of Gas and Electricity Markets
PFC	Power Finance Corporation
PLF	Plant Load Factor
PNGRB	Petroleum and Natural Gas Regulatory Board
PPA	Power Purchase Agreement
PSERC	Punjab State Electricity Regulatory Commission
PSPCL	Punjab State Power Corporation Limited
RfP	Request for Proposal
RfQ	Request for Qualification
SBD	Standard Bidding Documents
SBI	State Bank of India
SC	Supreme Court of India
SEB	State Electricity Board
SECL	South Eastern Coal Fields
SERC	State Electricity Regulatory Commission
SLC-LT	Standing Linkage Committee—Long Term
SPPA	Supplementary Power Purchase Agreement
TCE	Transaction Cost Economics
TPC	Tata Power Company
UMPP	Ultra-Mega Power Project

Chapter 1
Introduction

In Part A of this chapter, we introduce the rationale for regulating network utilities and trace the origins and evolution of two distinct models of regulation, one in the USA and the other, in Europe, the UK and the developing world. It then goes on to discuss the reorientation of this institution in the era of neoliberal reforms to subserve contractual relationships in the marketplace.[1]

In Part B, the evolution of electricity industry in India is discussed in detail to highlight the importance of contractual relationships in an unbundled industry and the salience of the adjudicatory role of the regulator in interpreting contracts.

1.1 Part A

1.1.1 Why Regulate?

For well over a century, unregulated market place was the norm. Economic regulation, generally understood as some form of intervention by the government or its agencies, has been around since the late eighteenth century (Priest 1993). Electric utilities were among the earliest industries to be regulated. Initially, in the USA, utilities like electricity and water were provided by municipal or local governments obviating the need for intervention, on the assumption that governments always act in the public interest. Later private investors joined the business, dominating the utility landscape in cities and towns. Edison Illuminating Company was set up in New York in 1880, followed by another in Chicago in 1882. Soon many cities in North America were being served by private electric and water supply companies. Initially, the proliferation of private utilities necessitated state intervention in some form,

[1] This chapter is derived substantially from the first author's successfully defended doctoral dissertation submitted to University of Technology, Sydney, in 2019.

S. Mahalingam and K. Mahalingam, *Independent Energy Regulation in a Developing Economy*, https://doi.org/10.1007/978-981-99-4398-2_1

because electricity lines, transmission grids, gas/water pipelines had to go through public and private lands. Subsequently, the rationale extended to protecting the public from price gouging.

While public interest is the implicit assumption in regulation, economists have been exploring other hypotheses such as whether regulation serves the public or private interests (Stigler 1971; Posner 1974), whether network utilities are necessarily natural monopolies (Demsetz 1976), whether the public interest can be served through instrumentality other than regulation, etc. Many of these economists went on to win the Nobel prize for their insightful analysis of the factors triggering regulation, deviating from the initially assumed 'public interest' argument.[2] Some of these theories will be discussed in detail in the next chapter.

Economic regulation acquired legitimacy in the rest of the world, in the post-Great Depression era when the modern 'welfare state' put down roots across the Western world and governments were expected to assume a welfare-enhancing role, providing basic services and amenities to their citizens. Roosevelt's New Deal in the USA was apiece with this trend. Post-war Europe took the welfare mandate quite seriously, rebuilding their war-torn economies and assuming the responsibility of providing a modicum of basic living standards to their populations such as provision of basic utilities.

In the USA, throughout the nineteenth century, American public utilities were typically operating at the municipal level. Between 1907 and 1922, investor owned utilities proliferated; nearly thirty states created 'independent' utility commissions to regulate them (Troesken 1996). The appointment of the Interstate Commerce Commission (ICC) in 1887 to regulate certain aspects of interstate railroad freight rates introduced independent regulation at the federal level. This role expanded over the twentieth century to cover telephone and telegraph as well. Subsequently, during the New Deal era, ICC was subsumed in the Federal Communications Commission (FCC) with a vastly expanded jurisdiction. The Federal Power Commission (FPC) came into being in 1920 and eventually gave way to the Federal Energy Regulatory Commission (FERC) in 1977 (Joskow 1985). Independent regulatory commissions in the USA are perpetual institutions intervening on all aspects of an industry's functioning and therefore likened to socialistic planning (Op cit. Priest 1993).

While US federal commissioners are appointed for a term, a few state commissioners are elected. Commissioners' independence derives from the manner in which they are insulated from political interference through security of tenure, financial autonomy, etc. Commissioners are appointed by the government on the basis of their political affiliation, although care has been taken in the structure of the commission itself to ensure bipartisan representation. The salience of the American regulatory

---

[2] For instance: Ronald H. Coase won the Nobel in 1981 for his contribution to the theory of property rights and its relevance for economic regulation. George J. Stigler won the Nobel in 1982 for postulating regulation is sought and obtained by the regulated entities themselves to retain their monopoly rents. Oliver E. Williamson won the prize in 2009 for his contribution to transaction cost theory.

commissions derives from its collegiate functioning. While commissioners in the USA may and do develop expertise on the mechanics and commercial aspects of the businesses they regulate, they are not necessarily selected for their domain expertise. Commissioners are called upon to balance the interests of consumers against those of investors. Herein lies the origins of 'rate of return' regulation where investors were assured a fair rate of return on their investment and constitutional protection against expropriation.

In the UK, regulation was exercised through government ownership of basic utilities supplying electricity, water, postal services, telecommunications, railroads, etc., obviating the need for a separate (independent) regulatory agency 'to protect' the public interest, the underlying assumption being democratically elected governments ignore the public interest at their own peril. Other continental governments either moved directly to more efficient state ownership or managed to achieve a better balance between public and private interests, often by encouraging cartels operating under loose regulation. Some evolved a more satisfactory and durable system of municipal ownership in partnership with the state for statewide networks, as with the Scandinavian electricity grids (Newbery 1999). Colonies in Asia and Africa too adopted the British or European model of regulation, mostly state/municipal ownership. Throughout the twentieth century, these two systems—one under state ownership, the other, in the USA under commission regulation, dominated the utility landscape worldwide.

1.1.2 The Era of Neoliberalism

In the last quarter of the last century, neoliberal economic policies inspired by what came to be known as the 'Washington Consensus' advocated superiority of markets over the state, in supplying all manner of goods and services, including basic utilities hitherto supplied by government-owned companies. Years of government ownership/control had eroded the business savvy of public corporations, seriously depleting their finances and jeopardizing the ability to replace ageing and creaky infrastructure, adding to the allure of the new ideology. With Margaret Thatcher in Britain and Ronald Reagan in the USA championing this new paradigm of governance, the idea gained currency rapidly, leading to structural changes economy-wide, in many parts of the world.

One such structural change involved the unbundling of hitherto vertically integrated utilities that produced, transmitted and supplied electricity/natural gas, etc., to retail consumers. The integrated electric utility was unbundled into its constituents—generation, transmission and retail distribution. Later, distribution was further unbundled into carriage and content or wires and supply businesses. In the UK, unbundling was followed by large-scale and often one-stroke privatization of electricity, water supply, natural gas, railroads, etc., and over the next few years, many OECD countries followed suit. France and a few provinces in other European countries remain exceptions, choosing to retain public ownership of their electric and gas utilities. What emerged in the 1970s and 1980s in Europe was a mixture of structurally unbundled utilities under diverse ownership regimes.

Such structural changes in network utilities called for an entirely new governance paradigm, more in the nature of 'deregulation'. The competitive segments of these unbundled industries were left to market forces (deregulation), while the monopoly segments like transmission networks and pipelines were to be handed over to the oversight of independent regulatory agencies. These agencies would also oversee introduction of competition in the sector by ensuring level playing field for all market players. In this domain, sectoral regulators share their turf with the competition regulator with a similar, but cross-sector mandate.

Subsequently, the integration of EU member states into a common market required standardization of national regulatory procedures and the institution of a supra-national EU-wide regulator in Brussels. In order to conform to the standards set by the EU regulator, all member states had to put in place independent regulatory agencies to regulate domestic network utilities.

Meanwhile, the IMF and World Bank assumed ownership of this new governance paradigm, nudging developing economies—erstwhile colonies of European powers—, to similarly restructure their utilities and set up independent regulatory agencies. In India, World Bank designed and implemented unbundling in the State of Orissa in 1996. Many developing countries unbundled their vertically integrated state-owned utilities into manageable segments, privatizing them where feasible and leaving them under state ownership where not. The unbundled utilities were also corporatized in the hope that they would now function like commercial businesses. Budgetary funding was withdrawn, and the unbundled utilities were left to fend for themselves in the marketplace.

At the core of these structural reforms and the new governance model is the independent regulatory agency (IRA), designed and overseen by experts from Fund-Bank institutions. The IRA is designed to be independent of all stakeholders, most crucially, the government. The relationship among stakeholders would henceforth be based on enforceable contracts.

The distancing of the democratic government from regulation necessitated some form of consumer protection. The view that regulation is designed to protect consumer interests is strengthened by the enactment of legislation in the UK, expressly mandating the regulator to champion consumer interests in the course of their functioning.[3] Even in the USA, commissions appoint representatives to argue on behalf of consumers. In developing countries, however, the regulator is positioned as an impartial arbiter among all stakeholders in the industry.

---

[3] https://committees.parliament.uk/committee/127/public-accounts-committee/news/98508/regulators-must-work-together-to-prevent-consumers-being-exploited/ accessed on 23rd May 2022.

The independence of the regulatory agencies is secured through a combination of design factors some of which are: the commissioners are appointed for a fixed tenure and the budget of the commission is not voted upon by parliament, but charged to the national fund. Regulatory accountability is ensured through a set of mechanisms which include a statutory framework within which regulators operate, parliamentary approval of regulations investing them with legal force as delegated legislation, policy framework of the government which is binding on the regulators and finally, through the power of judicial review of regulatory decisions and orders conferred upon the civil courts through a well-defined appeals process as well as, in some cases, domain courts with expert jurists. Other accountability mechanisms are established more by way of convention than by fiat such as adoption of transparent and participatory regulatory proceedings and speaking regulatory orders stating the rationale for decisions.

Post-1980, regulatory commissions proliferated across utility industries worldwide, giving the modern state the appellation 'regulatory state' (Majone 1994). The move from the welfare state to a regulatory state also implies a change in the quality of citizenship, from its political connotation to an economic one. For instance, in a welfare state, the citizen has a right to some basic minimum services, whereas, in the regulatory state, her status is reduced to that of a consumer, in other words, a contractual relationship with the state (Barron and Scott 1992).

In the UK and elsewhere in Europe where utilities have been successfully privatized, the key role of the regulator is market-making, ensuring non-discriminatory open access to transmission networks and gas pipelines to market players; the regulator also determines tariffs of transmission networks and grid pipelines for open access by competing producers and suppliers and checking anticompetitive behaviour by market players. Crucially, the regulator is tasked with enforcement of contractual obligations between stakeholders. Where there is competition, regulation becomes redundant. Only non-competitive segments of the industry like transmission networks or grid pipelines would be governed by contracts with the market players and, as such, need to be regulated for access and price. But more often than not, the regulator would monitor the market and has overall responsibility for its smooth functioning.

Within this regulatory model, there are minor variations. Members of federal regulatory commissions in the USA are appointed by the president, but are independent of the executive. At the time of appointment, they may belong to either political party, but care is taken to ensure fair representation of both national parties in the commissions. Some US states also have elected commissioners, but once elected, they are also independent of the executive. The US utility regulators license new infrastructure like electric lines and pipelines, on the principle of 'convenience and necessity' and determine tariffs where applicable, apart from their role as market makers. In the UK and elsewhere in Europe too, they have licensing and market-making mandates.

In developing countries, regulators are chosen from politically unaffiliated domain experts. The spread of infrastructure in developing countries being sketchy and less than comprehensive, regulators have a different and perhaps, far more onerous role.

Almost everywhere, structural reforms have been accompanied by withdrawal of budgetary funding for expansion/operation of network utilities. In the event, regulators are required to incentivize private capital to expand infrastructure to hitherto unserved populations. Besides, since outright privatization of utilities is not feasible in developing countries, regulators have to provide level playing field to new private entrants at the margin so that they can compete effectively against the incumbent government-owned utility which tends to dominate the market at least initially.

In other words, in the developing world, this has thrown up a hybrid model. Parts of the industry are government-owned and others, privatized. More than privatization of existing businesses, liberalization of the industry was the model adopted, i.e. retaining the state-owned incumbent, but allowing new private investors at the margin in the hope that in time, there would be enough players to create a competitive market. Besides, in an unbundled industry, relationships between the new private players and state-owned enterprises along the value chain were governed by contracts which needed to be enforced.

This hybrid model has created complexities for developing country regulators. Until such time competitive markets develop, regulators are expected to mimic markets. Developing country regulators have to also guard against conflict of interest where the government is both policymaker for the industry and a dominant player through its ownership of incumbent companies.

Yet, under Fund-Bank supervision, the role of the regulator in developing countries is largely modelled on the same principles that apply to developed countries where the reach of basic utility network infrastructure is almost universal. The one crucial difference is that in India, regulators have been vested with quasi-judicial powers to resolve disputes between stakeholders. In the USA, France and Germany, energy regulators are not tasked with dispute resolution functions. Stakeholder disputes are referred directly to separate administrative law judges in the USA. FERC has only regulatory responsibilities. The French counterpart, CRE (Commission de Regulation de l'Energie), does not have dispute settlement powers. The Dispute Settlement and Sanctions Committee, known by its French acronym 'CoRDiS' (Comité de règlement des différends et des sanctions), is tasked with settling disputes concerning access to and use of the public electricity and gas networks between operators and users, and also with penalizing infringements of the Energy Code. Similarly, the German energy regulator *Bundesnetzagentur für Elektrizität, Gas*, has other regulatory functions like licensing and tariff determination, implementation of non-discriminatory open access to networks, but not adjudicatory functions. The only exceptions are the UK where OFGEM, the energy regulator, has powers to determine disputes between entities regulated by it and the Australian Energy Regulator.

1.2 Part B

1.2.1 Electricity Regulation in India

Post-independence, India had adopted a federal political structure with a central government and 22 state governments largely organized as linguistically homogenous units.[4] It also adopted a mixed economy model where the public sector was assigned the commanding heights of the economy including those requiring extensive infrastructure networks such as basic utilities. Electricity, telecommunication, roads and highways, public transport including railways, water supply, sewerage, etc., came under public/municipal ownership. Defence production, steel and certain other strategic non-utility industries were also dominated by government ownership. Less strategic industries were left to the domain of domestic private enterprises which were also protected from international competition through tariffs.

The federal structure of the Indian Constitution assigned certain subjects (like defence, foreign policy, nuclear power) exclusively to the central government (Central List), while other areas of activity such as public order, public health, land, water, etc., were entrusted exclusively to the State governments (State List). A few areas not in either Central or State List were assigned concurrent jurisdiction for both the central and state governments (Concurrent List). Electricity was assigned to the Concurrent List of the Indian Constitution. Electric Supply Act 1948 organized electric supply as a vertically integrated operation at the level of states. The State Electricity Boards (SEB) and, in some cases, departments of the state governments were responsible for generating, transmitting and supplying electricity to households and industries as well as for street lighting, etc. A few private integrated electric utilities supplying cities, like Bombay, Calcutta, Ahmedabad, Surat and Baroda prior to Independence, were allowed to continue.

Long-term power planning was the responsibility of Central Electricity Authority (CEA) which, in consultation with SEBs, drew up plans for capacity expansion. Capacity addition was executed by SEBs usually with financial assistance from state governments or through central plan grants. However, in most states capacity addition did not keep pace with the growing electricity demand, largely due to limitation of government finances and inadequate revenue of SEBs. Therefore, in 1975, Government of India established two wholly owned PSUs, National Thermal Power Corporation (NTPC) to build thermal generating stations and National Hydro Power Corporation (NHPC) to build large hydel projects; it was envisaged that power generated from these would be supplied to more than one state. Nuclear Power Corporation (NPC) was established in 1987 to add nuclear power to the energy mix. Powergrid, a transmission company to transmit bulk power to state grids was set up in 1989. Thus, central intervention in India's power sector was in the form of large thermal, hydel

---

[4] Now there are 28 states and eight union territories.

or nuclear power plants each of which would supply bulk power to more than one state. Naturally, the quantum of allocation of power from a central generating plant was decided by NTPC/NHPC/NPC, through mutual consultation with the concerned state governments and was executed through long-term power purchase contracts with the SEBs of recipient states.

In 1991, India decided to adopt structural overhaul of her economy, persuaded more by fiscal compulsions than by the conviction that markets would deliver better efficiencies than state ownership of core and strategic industries. Naturally, power sector was one of the foremost focus areas for reform. India's state-owned power utilities, enfeebled by decades of populist policies, were in a parlous state, their finances precarious, bloated with surplus staff, plagued by declining efficiency and lackadaisical in catering to the burgeoning demand for electricity in a country with a rapidly expanding population.

It, therefore, seemed logical that reforms should first target adding new generation capacity to counter recurrent and growing shortages. With the governments at the centre and the states constrained for resources, the capital for augmenting capacity would have to come from private sector. Therefore, the initial reform measures were designed to incentivize private capital to build new generating capacity. This was done through a new policy.

The Independent Power Producer (IPP) policy promulgated in 1992 was broadly in line with policies adopted elsewhere in the developing world facing similar shortages. It envisaged a 'two-part tariff' to be paid to any new power plant developer—domestic as well as foreign. The 'fixed' part of the tariff would compensate the power plant developer for investing in the power plant in the first place and has no reference to actual power generation, the only requirement being the plant should always be available to supply 85% of its capacity (also called PLF or plant load factor). Fixed part of tariff comprised elements such as 16% post-tax return on equity, interest on debt, and depreciation at predetermined rates, all payable over a long period not exceeding 30 years, discounted to net present value and levelized. The variable part of the tariff would be directly related to actual power generation and comprised fuel and O&M costs which were a pass-through in the tariff. The split in the tariff design was to provide a modicum of comfort to the developer by assuring a revenue stream to take care of the cost of building the plant regardless of whether the utility was in a position to absorb the power it generated. In addition, the developer would be compensated for fuel and O&M costs on actual basis, apart from certain bonuses linked to performance above-specified levels.

The buyers of such power would be the vertically integrated SEBs. This was referred to as the 'single buyer model' wherein the State Electricity Boards signed 30-year contract with the power developer to purchase power generated by the IPP and pay all fixed costs over 30 years through the fixed part of the tariff; variable costs would be paid on actual generation.

The initial policy envisaged awarding power purchase contracts (PPA) to IPPs on Memorandum of Understanding (MoU) basis. Under this arrangement, the power developer would first enter into an MoU with the SEB, build the plant and then present the costs to the SEB for fixation of tariff. In other words, these were referred to as cost-plus projects with potential problems of cost-padding and gold-plating associated with such an arrangement. In the next few years, hundreds of MoUs were signed by many state governments with private investors, but few projects actually materialized (Prayas 2011).

The MoU paradigm was conceived when India was uncharted territory as far as foreign (and even domestic) investors were concerned and would therefore be reluctant to compete for projects in a competitive bidding process, India being an unfamiliar economic terrain. In addition, a limited number of prospective developers of power plants were also offered payment security in the form of sovereign guarantees. Under this arrangement, even though the SEBs signed the PPAs, the annual payment of tariffs to the developer would be guaranteed by the respective state government and counter-guaranteed by GoI. If the SEB failed to pay the developer, the state government would pay, and in the event of the state government failing to do so, the centre would step in to pay the IPP and deduct the sum from the respective state's share of taxes. Enron Corporation was the first IPP project to be set up in Maharashtra, and it almost bankrupted the Maharashtra government thanks to such guarantee and after the Enron debacle sovereign guarantees were discontinued (Mehta 2000).

Whether MoU-based or commissioned through international competitive bidding, contracts became the fulcrum on which the commercial relationships would rest.

In 1998, The Electricity Regulatory Commissions Act was passed by the central government, and Central Electricity Regulatory Commission (CERC) was set up to regulate both central sector power projects and IPPs supplying to more than one state. Section 62 of Electricity Act 2003 authorizes the regulator to scrutinize the capital cost of projects and determine the tariff of power produced by the central generating stations and IPPs supplying more than one state; it also determines the transmission tariff of Powergrid and regulates the transmission and access arrangements between stakeholders. Apart from tariff determination, CERC has also been endowed with quasi-judicial powers to adjudicate on disputes between stakeholders in the industry under its jurisdiction. In its adjudicatory role, CERC has been empowered to settle disputes between generators and discoms.

The 1998 Act also authorized the setting up of State Electricity Regulatory Commissions (SERC) to regulate the state electric utilities, whether vertically integrated or unbundled. The main functions assigned to state regulators include determining the power purchase cost of discoms through a merit order dispatch system wherein the cheapest power is dispatched first, followed by the next cheapest and so on until all demand is met. But the regulator is not required to actually scrutinize generation costs except in the case of new cost-plus plants for which regulatory approval was sought. SERCs also determine the intra-state transmission and distribution tariffs and the final consumer tariffs.

Orissa ERC had been set up in 1996 under the supervision of World Bank, and the same design was used in the other SERCs set-up after the ERC 1998 Act was passed. Many states chose to unbundle their SEBs into generation, transmission and distribution segments, although the unbundled entities largely remained under state ownership. Many of the unbundled entities across the country were also corporatized, giving them a commercial persona. At this stage, India had a mix of vertically integrated utilities whose tariffs were determined by their respective governments as well as unbundled and corporatized utilities purportedly fixing tariffs based on commercial considerations.

While corporatization of the unbundled distribution, transmission and generation businesses was accomplished without much difficulty, privatization met with limited success. A few state-owned generation plants such as Ib Valley power plant in Orissa were privatized successfully, but most others remained under state ownership. Privatization of discoms remained a non-starter except in the State of Orissa and Delhi. In Orissa, privatization of discoms hit a roadblock and reverted to the government ownership in 1999. In 2020, once again, these discoms were privatized. In Delhi, on the other hand, with determined government support, all three discoms were successfully privatized in 2002 and continue to function as private utilities.[5] The historically private discoms in Kolkata, Ahmedabad, Surat, Bombay, Baroda and Noida continue to remain in private ownership.

Structural reforms in the power sector entailed a paradigm shift in the way unbundled discoms sourced power—through PPAs with generators—state-owned, central or IPP. Unlike in the UK, there was no wholesale day ahead power market in India. Discoms also had to enter into transmission agreements with transmission utility to transport the electricity to the bus-bar (city gate) of their respective franchise areas. Thus, PPAs came to be the only mode by which discoms could purchase power directly from generators except for a small supplement from real-time markets to tide over exigencies. In the event, the relationship between the unbundled entities came to be governed by contracts.

1.2.2 Contracting Power Purchase Through Competitive Bidding

Meanwhile, MoU-based power development had been replaced by competitively bid power projects. The successful bidder satisfying prespecified technical criteria would be selected on the basis of the lowest composite tariff (fixed plus variable tariff) it would charge for generating power from the proposed power plant over the next thirty years. In 1995, India's Ministry of Power, exercising its concurrent jurisdiction, issued guidelines for conducting competitive bids for selecting the developer for setting up power plants.

---

[5] Orissa discoms have once again been privatized through the PPP model in 2020.

In 2003, Government of India overhauled and harmonized its existing electricity laws and put in place a comprehensive legislation that would govern the industry. Electricity Act 2003 furthered the reforms initiated in 1992. It mandated all state governments to set up independent regulatory commissions to fix consumer tariffs and regulate the sector, signalling centralization of control over electricity, a concurrent subject under the Indian Constitution.

In 2005, a revised guideline was issued for competitive bidding for power generation projects, and this was binding on state discoms. The stated objective of the new guidelines was to "promote competition in procurement of power by distribution licensees; facilitate transparency and fairness in such procurement; reduce the information asymmetries for various bidders and encourage competition; expedite the process of materialization of projects; and provide flexibility to suppliers in managing internal operations, while ensuring the availability of power at predictable rates and thus protecting consumer interests" (GoI guidelines, 2005 Annexure 1).

The bid document specified two ways of bidding: Case 1 where the location, technology or fuel is not specified by the procurer discom, and hence, the developer has full freedom to decide these. Case 2 is location and fuel-specific bidding, i.e. the procurer specifies the location and/or fuel and is also responsible for arranging of the same. Similarly, large hydro power plants and load centre thermal projects also come under this category. MoP also notifies from time to time, the Standard Bidding Documents (SBD) such as the Request for Qualification (RfQ), Request for Proposal (RfP) and Power Purchase Agreements (PPA) for both Case 1 and Case 2.

For instance, some bidders might prefer to pass on escalation in the cost of fuel to the tariff, deeming it beyond their control, while others might be willing to risk a uniform fuel cost and O&M (variable cost) through the life of the project. Therefore, GoI guidelines allow for both the scenarios. A power project developer could choose ex-ante, whether to go with Case 1 or Case 2, the latter allowing fuel cost changes to pass through to the tariff, while the former entailed either a uniform tariff for the entire life of the project or an escalating tariff with the quantum of escalation already determined and frozen ab initio in the PPA. There is also a provision for bidding part frozen and part escalable tariffs. Once the developer chose to bid under a particular scenario and won the bid on that basis, it would have to honour the terms of the contract.

The following flowchart lists the various processes and institutions involved in approving competitively bid power projects:

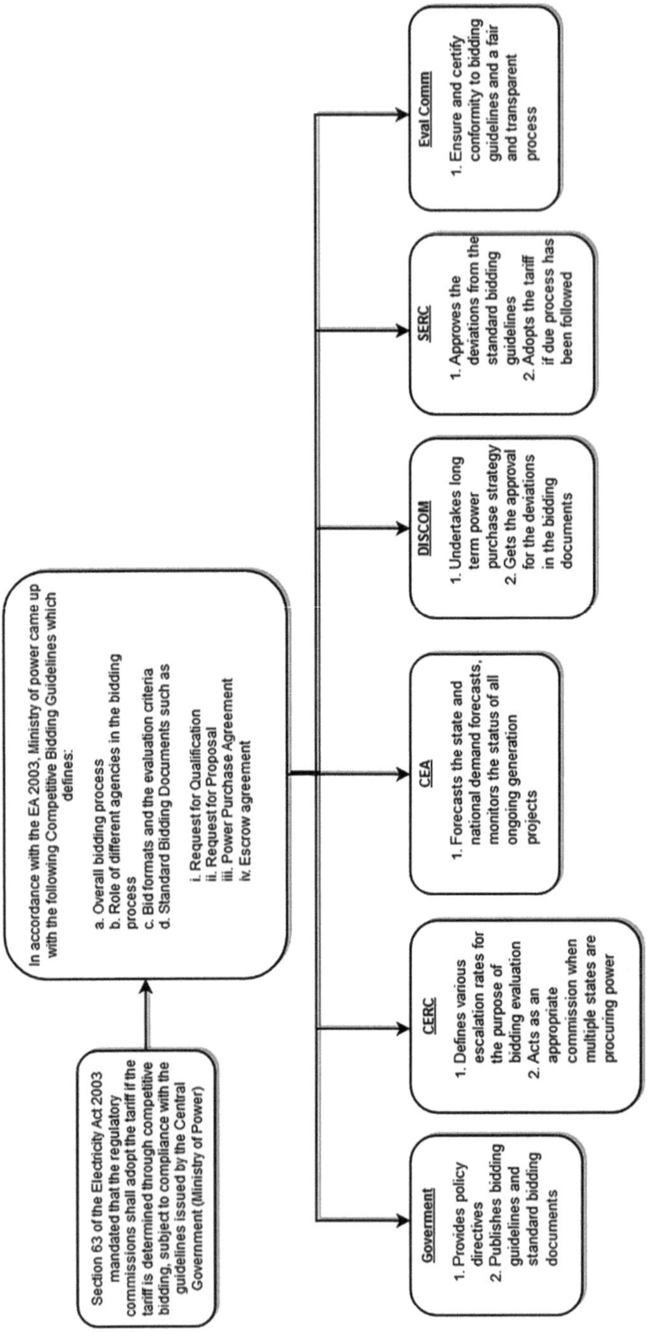

Source Prayas Energy Group (2011) (permission obtained)

1.2.3 Role of India's Electricity Regulators

Section 62 of Electricity Act 2003 mentions that the determination of tariff for the supply of electricity by a generating company to a distribution company is under the jurisdiction of the Electricity Regulatory Commission. However, section 63 (Determination of tariff by bidding process) states "Notwithstanding anything contained in section 62, the Appropriate Commission shall adopt the tariff if such tariff has been determined through transparent process of bidding in accordance with the guidelines issued by the Central Government". Thus, the regulator's remit in scrutiny of PPA is circumscribed in the case of projects awarded through a competitive bidding process conforming to GoI guidelines.

In a structurally reformed power sector governed by contractual relationships between the constituents and one where supply of electricity is predicated upon coordination between the constituents, sanctity of contracts assumes paramount importance. In a market economy, sellers and buyers willingly come together to conduct mutually beneficial transactions through legally enforceable contracts. Failure to perform the contract by any link in the chain leads to collapse of the market. Failure attracts specified penalties usually agreed upon in the contract itself.

As already stated, Electricity Act 2003 pre-empts the regulator's role in ex-ante scrutiny of PPAs of competitively bid projects on the assumption that markets will always throw up optimal least cost solutions. Ironically, however, ex-poste, the same regulator could be called upon to scrutinize the PPAs of competitively bid projects in their adjudicatory capacity when a complaint is filed by one of the contracting parties. Most disputes relate to the contractual arrangements between the utility and the power generator. Both CERC and SERCs are inundated with disputes which now take up a substantial chunk of regulatory time and resources, while their main task—of regulating the sector—has receded to an annual ritual largely outsourced to consultants.

Regulators are the first port of call for disputing stakeholders in the power sector. In recent years, most SERCs and CERC had to deal with numerous disputes related to contracts. While some of them are escalated to APTEL and yet further to the Supreme Court of India, judging by the number of disputes being adjudicated by regulators, many of them obtain closure at the regulatory level. Disputes relating to large projects, those that involve not only large sums of money but also substantial questions of law, are escalated right up to the Supreme Court.

This research focuses on adjudicatory functions of India's energy regulators, especially those that entail interpretation of contracts between stakeholders. Over the last quarter of a century, all energy regulatory commissions—CERC, SERCs and over the past fifteen years, India's natural gas regulator PNGRB—have established fairly standardized processes for their conducting their regulatory functions. There have also been a few scholarly studies that assess the effectiveness of regulators in their regulatory space. But so far, there has been no scholarly inquiry into the quasi-judicial functions of India's energy regulators.

Independent regulators were installed to oversee the transition of the energy sector from state ownership to a market-driven industry; regulators would have to provide level playing field to new market entrants competing against incumbent government-owned monopolies. They would also oversee the contracting relationships between the unbundled entities—some private and others, still government-owned—to ensure smooth functioning of the sector. They have the tricky task of adjudicating disputes between government-owned distribution and transmission companies and private generators. Adjudication of disputes calls for not only interpretation of statutes and contracts, but also creates a template for the evolution of jurisprudence appropriate to a new economic environment, one in which both government and private players are required to work harmoniously with each other to deliver an indispensable and essential commodity like electricity. Therefore, adjudicatory orders of regulators in their quasi-judicial role provide a good reflection of regulators' expertise and appreciation of the importance of water-tight commercial contracts as well as their own role in such interpretation in the new economic context.

The case studies in this book have been chosen for their relevance in shaping jurisprudence in a regulated industry with a hybrid market structure. The disputes analysed in this research span a spectrum of issues salient to the success of reforms in a liberalizing economy. They include

(i) sanctity of contracts;
(ii) interpretation of contracts;
(iii) implying terms into contracts;
(iv) credible commitment to reforms;
(v) competition in retail supply;
(vi) turf wars in the regulatory paradigm.

Apart from deconstructing regulators' quasi-judicial orders to identify their perspectives on a liberalizing industry, this study also attempts to examine the Indian experience against extant regulatory theories. Crucially, the study seeks to find out if the hybrid structure of India's electricity industry poses a challenge to understanding the Indian experience against the backdrop of the existing theoretical framework.

1.3 Chapter Scheme

Chapter 2 briefly examines the various theories of regulation including public interest theory, private interest theory, capture theory, etc., to provide a theoretical context and framework to this research.[6] In analysing the transformation of regulation in the era of neoliberalism, this chapter highlights the salience of the role of regulator as a surrogate for markets based on contractual relationships.

[6] This chapter draws substantially from the first author's successfully defended doctoral dissertation submitted to University of Technology, Sydney, in 2019.

Chapter 3 tracks three landmark disputes first adjudicated by regulators, but escalated to APTEL and thence to the Supreme Court, through their tortuous legal labyrinth lasting over 17 years and have still not obtained final closure. This chapter speaks to the core hypothesis of this book, namely (i) that competitive markets are easier conceived than realized in developing economies with their hybrid market structures—part government-part private, (ii) that when the contracting parties are unequal, thanks to the asymmetry of relevant and critical information available to either of them, competition turns out to be a chimera to the detriment of one party, (iii) long-term time-consistent contracts are difficult to write and are fraught with such complexities that they carry within them, seeds of their own supersession, and (iv) that jurisprudence in a liberalized economy treads a chequered evolutionary pathway, informed by the diverse and at times conflicting perspectives of an array of stakeholders.

Chapter 4 examines a singular case study—of an MoU-based project that signed a contract with APSEB in 1998, but it was commissioned only in 2016. The obstacles faced by this project were partly due to lack of credible commitments on the part of governments and their entities that signed the PPA as well as weak commitment on the part of the project developers. The hapless regulator was caught in this tussle where none of the stakeholders seemed to be committed to the contract they signed, leaving the consumers in the lurch.

Chapter 5 examines three case studies that involve interpretation of contracts by regulators, APTEL and the Supreme Court of India; it highlights the similarities and differences in their perspectives and the consequences thereof when the cases obtain res judicata at the level of the Supreme Court. This chapter specifically looks at disputes raised on the basis of perceived ambiguity in the drafting of the contract, and the responses thereof, of the three adjudicatory authorities. It also touches upon the consequences of such divergent perspectives of key stakeholders for India's electric supply industry.

Chapter 6 follows the rambling and complex path taken in the implementation of a provision in the law that mandates the regulator to introduce competition in a traditional monopoly business such as electricity distribution that too in a large city like Mumbai. It examines how various stakeholders responded to the implementation of this provision and the complexities it created for the regulator. This chapter highlights the superiority of expert adjudication in dealing with specificities of the electricity market.

Chapter 7 deals with three core issues of paramount importance to the success of the regulatory regime. First, it analyses the difficulties encountered by an independent regulator while regulating a state-owned utility. A legal challenge to the regulator's remit posed by a regulated government utility throws up in sharp relief, the contrasts between the perceived legitimacy of the government vis-a-vis that of the regulator and the consequences thereof, especially in the context of lack of government commitment to reforms. It highlights the complexity of regulating in a hybrid system. Second, this chapter also delves into problems thrown up by acute information asymmetry between the regulator and the regulated entity and the contradictions that ensue as a consequence; it highlights the differences in the understanding and appreciation of the role of the regulator in a liberalized economy, especially the variance between the statutory intent and judicial interpretation. It points out how jurisprudence in this case has introduced distortions into the governance paradigm. Third, this chapter also throws light on the turf wars that vitiate the business environment when liberalization of the industry has evolved not in response to a grassroot need, but to a top-down imposition without adequate buy-in from the government of the day.

Chapter 8 is the concluding chapter containing summary, findings and policy implications for improving the regulatory paradigm. The first part of chapter summarizes the findings from the study, the conclusions therefrom and the policy implications thereof. The second part of the chapter examines the Indian regulatory performance in the light of theories of regulation discussed in Chap. 2 to identify areas in a developing country context that defy conformity with such theories.

References

Barron, Anne, and Colin Scott. 1992. The Citizen's Charter Programme. *The Modern Law Review* 55 (4): 526–546.

Demsetz, Harold. 1976. Economics as a Guide to Anti-trust Regulation. *Journal of Law and Economics* 19 (2): 371–384.

Joskow, Paul. 1985. Long Term Vertical Relationships and the Study of Industrial Organization and Government Regulation. *Journal of Institutional and Theoretical Economics* 586–593.

Majone, Giandomenico. 1994. The Rise of the Regulatory State in Europe. *West European Politics* 17 (3): 77–101. https://doi.org/10.1080/01402389408425031.

Mehta, Abhay. 2000. *Power Play—A Study of the Enron Project*. Orient Blackswan.

Newbery, David M. 1999. *Privatization, Restructuring, and Regulation of Network Industries*. Cambridge, MA, London.

Posner, Richard A. 1974. Theories of Economic Regulation. *Bell Journal of Economics and Management Science* 5: 335–358.

Prayas Energy Group. 2011. Prayas Discussion Paper on Capacity Addition Through Competitive Bidding in India Private Sector Participation in Generation: From MoUs to Competitive Bidding, March 2011.

Priest, G.L. 1993. The Origins of Utility Regulation and the "Theories of Regulation" Debate. *The Journal of Law & Economics* 36 (1), Part 2. In *John M. Olin Centennial Conference in Law and Economics at the University of Chicago*, April 1993, 289–323. The University of Chicago Press.

Stigler, George J. 1971. The Theory of Economic Regulation. *Bell Journal of Economics and Management Science* 2: 3–21.

Troesken, Werner. 1996. *Why Regulate Utilities?: The New Institutional Economics and the Chicago Gas Industry, 1849–1924*, 132 pp. Ann Arbor: University of Michigan Press.

Chapter 2
Theories of Regulation

In this chapter, a brief outline of the theoretical underpinnings of economic regulation as it evolved historically has been attempted. Public interest and private interest theories as well as transaction cost theories are discussed. The chapter then goes on to trace the advent of neoliberal economic theory advocating free markets in the provision of utility services and the attendant deregulation. The discussion expounds on the assumptions underlying the institution of independent regulatory agencies in developing countries.

2.1 Public Interest Theory

In this section, we discuss theories of regulation that emerged over the past hundred years, providing perspectives on and justification for, utility regulation. An understanding of the theoretical framework of regulation is essential to draw up the template for the main research question this book purports to address, namely, the perspectives of key stakeholders on the role of regulation in a liberalized economy.

For over a hundred years, regulation was justified on grounds of public interest. This arose from the monopoly nature of network infrastructure required to supply utilities like water, electricity, cooking gas, etc. Since duplication of infrastructure is infructuous, and scale economies, substantial, the monopolist builder and operator of infrastructure could game the market, pricing the products above marginal cost, curtailing production below optimal levels leading to deadweight loss in the economy.[1] Economic theory emphasizes that price must equal marginal cost in an allocatively efficient market. Where price does not equal marginal cost, intervention is warranted to prevent or at least, minimize, deadweight losses. Hence the need for regulation.

[1] Deadweight loss denotes the loss of output, and therefore, welfare, in a monopoly market where prices are above marginal cost and output is limited to maximise profit.

S. Mahalingam and K. Mahalingam, *Independent Energy Regulation in a Developing Economy*, https://doi.org/10.1007/978-981-99-4398-2_2

Regulation is a product of "responses of government to public demands for the rectification of palpable and remediable inefficiencies and inequities in the operation of the free market" (Posner 1974). Among the proponents of this theory are also Bonbright (1961), an economist. This theory argues that there are inefficiencies and inequities in the market place and that regulation is supplied in response to the demand of the public for correction of such inefficiencies and inequities.

The proponents of this theory argue that behind each specific instance of regulation (viz. worker protections, subsidies, price regulation for utilities), there is a specific market imperfection that "... supplie(s) a complete justification (for the regulation)" (Posner 1974). Empirical evidence provides little support for this theory, as little correlation has been found between the extent of regulation in an industry and how prone it is to market failures in the form of externalities and undue concentration of market power. Crucially, the evidence also provides little correlation between regulation and increases in efficiency, welfare, or equity, as we shall see later in this inquiry.

Public interest theory introduces the concept of regulation for market failure on multiple counts. It could arise from any one of the following:

(i) discrepancies between costs and benefits arising out of imperfect competition;
(ii) unstable markets;
(iii) missing markets; or
(iv) undesirable market results.
The first arises out of imperfect competition which will cause prices to deviate from marginal cost. When prices are lower than marginal cost, the supplier may be unwilling to supply the goods/services, leaving out those segments of the population who are responsible for such deviation. This is the deadweight loss that is avoided by regulation.

Unstable markets are characterized by dynamic inefficiencies with respect to the speed at which these markets clear or stabilize. These instabilities waste scarce resources. Missing markets imply the demand for socially valuable goods and services for which total value exceeds total cost but where prices or markets do not arise, such as in the case of externalities or markets with particularly high transaction costs. And finally, even if the competitive market mechanism allocates scarce resources efficiently, the outcomes of the market processes might still be considered unjust or undesirable from other social perspectives. Hence, the government steps in to act as an omniscient, omnipotent and benevolent regulator (Dixit 1996).

2.1.1 The Imperative of Universal Supply Argument

The public interest argument runs thus: the fact that essential utility services need to be made available to everyone in society, on demand, on a non-exclusionary basis, necessitating universal supply implies the need for extensive and comprehensive network infrastructure to reach every potential consumer. Substantial upfront investments are required to set up such networks, yet utility services are typically consumed in low volumes, universally or by a large segment of the population. Furthermore, as this trickle of demand is unlikely to expand exponentially, recouping just the cost of the initial investment entails long recovery periods for the investors in network infrastructure. Electricity, urban water supply and sanitation, roads, telecommunications and piped cooking gas fall in this category. Left entirely to market forces, network infrastructure might cluster around relatively affluent urban population centres where investment would both be lower, and cheaper, due to ease of access and supply, as well as clustering of demand over relatively small geographical areas. Investment would consequently yield higher returns, even as it leaves out rural and far flung consumers as well as the urban poor. Some kind of regulatory intervention would be required in the larger public interest, to ensure that everyone in society has equitable access to utility services.

The problem isn't just one of supply, though that certainly is the most difficult to tackle. The situation is complicated by the fact that public interest, in the context of a welfare state, also requires that utility services must be supplied to everyone in society regardless of their ability to pay. For instance, in cold countries, regulators often decree that heating not be turned off in winter to poor consumers who are unable to pay. Even in the USA, FERC decrees that electricity or heating not be turned off for non-paying consumers. In fact, most utility services have become so indispensable to modern life that a life of dignity is inconceivable without them. This, then, is the problem—the failure of the market to provide both universal *access* in addition to universal and continuous *supply* of utilities society deems essential to human life and dignity which public interest regulation aims to fix.

2.1.2 The Political Imperative Argument

This is not to say that the output of utilities should be termed 'public goods' as defined by classical economists; they are not. Yet, treating them at par with other private goods to be supplied by markets to those who can afford to pay for them poses a problem, especially in the context of the modern welfare state. Democratic governments are typically reluctant to risk failures in provision of basic utilities, for self-preservation if nothing else. Even non-democratic governments need a degree of legitimacy to avoid being overthrown and hence need to ensure universal access to utility services.

In fact, 'public interest', often used to justify government or regulatory intervention is too vague a term and too fragile a basis for practical regulation (Feintuck 2010). It has been argued that precisely because it is difficult to identify or formulate what constitutes public interest, regulation is 'relegated' to serve measurable economic, rather than social or collective interests.

2.1.3 The Natural Monopoly Argument

Most basic utilities need infrastructure networks to supply and hence tend to be natural monopolies with substantial scale economies. This is because construction of the requisite infrastructure, which by itself provides no revenue, cannot be decoupled from transport or supply. After construction, the owner of the infrastructure can either allow others to supply the utility, serving solely as transport, or might choose to supply the utility itself. In either case, such a natural monopoly can and often would extract monopoly rents, forcing up prices and leading to consumer exclusion. Of course, on the other side, there is the 'hold-up' problem where stakeholders, especially the government, can renege on payment for infrastructure once it has been built. Regulation is a way of keeping a check on such abuses. Economic regulation is primarily aimed at ensuring the monopoly firm does not charge prices far above marginal cost or limit output of an essential commodity or service. Viewed from this perspective, the objective of regulation would be to improve allocative efficiency and, thereby, welfare.

There are other public interest justifications as well, for the introduction of regulation. The government might intervene in the market so as to protect consumers' interests arising out of information asymmetry between producers and consumers (Breyer 1984). After all, a well-functioning market assumes perfect and complete information on the part of buyers and sellers, something virtually non-existent. In fact, in utility industries, the firm has all the information on costs, whereas the consumer has none. An empowered regulator can demand and access relevant cost and other information from the utility for the benefit of the consumers.

2.2 Capture/Private Interest Theory of Regulation

The public interest theory of regulation assumes a benevolent government, an assumption that has been questioned by many detractors. Capture theory, espoused by the Chicago School, emerged to counter the public interest theorists. According to this school of thought, the outstanding proponents of which are Posner (1971, 1974), and Peltzman (1976), regulation is a device for transferring income to well-organized groups, if the groups will return the favour with votes and/or contributions to politicians. Governments, like other stakeholders, also expect to benefit from regulation in the form of political support. Whether such support comes from the electorate (consumers) or from interest groups which fund elections, depends on the individual contexts and circumstances.

2.2.1 Stigler's Interest Group Theory

Stigler, on the other hand, points out that regulation is often sought and procured by the regulated monopoly itself in order to block competition (Stigler 1971). His hypothesis, for which he was awarded the Nobel prize in 1982, was that regulation is secured by politically effective interest groups, invariably producers rather than consumers. According to him, the central tasks of the theory of economic regulation are to explain who will receive the benefits or burdens of regulation, what form regulation will take and the effects of regulation upon allocation of resources. Stigler's theory predicts that regulation is likely to restrict entry as the incumbents have more bargaining power than potential entrants, and their rents will be threatened by competition.

From the foregoing discussion, it is evident that distributive goals, fairness or justice seldom figure in the scheme of economists expounding on the capture theory of regulation although governments might use these terms to justify introduction of regulation.

2.2.2 Posner's Disguised Taxation Theory

Posner goes one step further and posits that consumers could be co-opted into endorsing regulation. If entry of a new firm into an existing market increases deadweight losses and makes negotiations costlier, even consumers could be persuaded to oppose competitive entry by offering them cross-subsidies (Posner 1971) or targeting benefits appropriately (Crandall 1983). Not only can interest groups capture regulation, but can even make accomplices of out of consumers whose interest regulation is meant to serve, in the first place.

Posner further argues that regulation is a form of disguised taxation; that is, through regulation, government attempts to tax one group for the benefit of another group. As with Stigler, Posner's theory of regulation is a theory of income redistribution which ties in with government capture of the institution for political purposes. The inefficiency of regulation emerges as a consequence of the selective tax system employed. In the Peltzman model, the regulator allocates benefits from regulatory policies among different groups to maximize political support (Peltzman 1976). Other economists believe governments seek regulation to extend their tax base for rent maximization (North 1990).

2.2.3 Bureaucratic Theory of Regulation

Bureaucracies capture regulation for their own purposes. Scholars McKenzie and Macaulay point that regulation is purposefully designed by public bureaucracies to make the private sector inefficient, increasing the relative attractiveness of the public sector. The consequence of regulation may be 'economic waste', but regulation can nevertheless be an important tool of bureaucracies interested in expanding the size and scope of their own economic realms. Akin to the above is the private interest theory which predicts that regulators will use their power to transfer income from those with less political power to those with more (McKenzie and Macaulay 1980).

From the discussion above, it is evident that economic regulation is the outcome of bargaining by various interest groups, each seeking to maximize its own utility. Regulators are seen as utility-maximizing arbiters between these various interest groups. Different interest groups have different bargaining power, depending upon their costs of organizing and the benefits of manipulating outcomes, and they will intervene to redistribute benefits to their group at some additional inefficiency cost. Just as the central insight in public finance is that redistribution creates deadweight costs that must be balanced against the distributional gains of taxation, so interest group theory recognizes that policy interventions will be biased in favour of some groups at the expense of others and at some cost in extra deadweight losses. The outcome may bear little relationship to the best feasible option that a benevolent dictator might choose in the public interest and certainly need not minimize efficiency losses.

2.3 Institutional Economics and Transaction Cost Theories

In the 1970s, New Institutional Economics emerged as a distinct discipline to explain types of and the rationale for industrial organizations. The seeds of this new field of inquiry were sown by Ronald Coase in 1960. Coase maintained that regulation was necessitated by the presence of transaction costs which arose because property rights were not clearly defined. Any issue that arises as or can be reformulated as a contracting problem is usefully examined through the lens of transaction cost economizing, according to Coase who went on to win the Nobel for his theorem. In other words, where there is a conflict of property rights, the involved parties can bargain or negotiate terms that are more beneficial to both parties than the outcome of any assigned property rights. It also asserts that in order for this to occur, bargaining must be costless; if there are costs associated with bargaining, such as those relating to meetings or enforcement, it affects the outcome. The Coase Theorem shows that, where property rights are concerned, involved parties do not necessarily consider how the property rights are granted if they can trade to produce a mutually advantageous outcome (Coase 1960).

2.3.1 *Williamson's Bounded Rationality Theory*

Oliver E. Williamson carried this concept forward to explain different kinds of market behaviour, based on bounded rationality which takes the study of regulation to the sphere of behavioural economics (Williamson 1981). Bounded rationality, opportunism, the primacy of markets and the action of economizing are building blocks of transaction cost economics, as formulated in the work of Williamson. Bounded rationality refers to the limits on the ability of an individual to acquire, assimilate and use information and anticipate all future contingencies in a contracting relationship. Opportunism assumes that some individuals will take advantage of information asymmetries, not be forthright in their dealings, etc. Building on those elements, TCE has become a new orthodoxy among contemporary theories of the firm (Groenewegen and Vromen 1996). It has been further developed by other economists such as Alchian (1961) and Demsetz (1967).

2.4 Part 3—Theories of Independent Regulation

2.4.1 *Credible Commitments, Expert Regulators, Rule-Based Regulation and Time-Consistent Contracts*

In the earlier part, we dealt with theories of regulation that did not distinguish between government or independent regulation, but in this section, we look at the triggers and origins for independent regulation. Joskow defines independent regulation as "a quasi-judicial structure that applied transparent administrative procedures to establish prices, review investment and financing plans, and to specify and monitor other terms and conditions of service" (Joskow 2007). But this definition does not explain why regulation should be independent or non-majoritarian, unlike the government which establishes it in the first place.

Why do non-majoritarian institutions, which by design are not directly accountable to the voters or their elected representatives, emerge at all in democratic policies where public policy is supposed to be made by electorally accountable people? According to Majone, the cost of operating the political process and, in particular, the cost of reaching and enforcing political agreements may be the reason why political principals choose to delegate policymaking powers to non-majoritarian institutions like independent central banks, regulatory bodies or international organizations, rather than making policy themselves. However, democratic politicians delegate powers in some areas but not in others: they may delegate responsibility for monetary policy to an independent central bank, but they continue to make tax policy themselves, despite the demands of time and expertise that this entails. "The transaction-cost approach suggests that democratic politicians will prefer to make policy themselves as long as the political benefits they derive from doing so outweigh the political costs; otherwise, they will delegate powers" (Majone 1994).

Democratic governments eyeing the next elections tend to make short-term policies, also referred to as 'time-inconsistent' policies (Laffont and Tirole 1993). Delegation of powers to an administrative agent will not, in general, be sufficient to resolve the government's time-inconsistency problem. Only an independent delegate, such as an independent regulator, not subject to the power of direction of the delegating authority, can provide credibility to long-term policy commitment.[2] Therefore, the relationship between the government that creates the regulatory institution and the regulatory agency cannot be defined within a principal-agent framework. For, if the regulator is an agent of its principal, namely the government, it will have to conform to the latter's will and directions in every aspect, whereas the government sets up an independent agency precisely to execute politically inconvenient but time-consistent policies which it is not itself in a position to do.

Independent regulation is also ineluctably linked to private provision of utility services. In democratic societies even in a neoliberal context, the government is the last resort for citizens seeking satisfactory provision of basic utilities without which modern life is inconceivable. Democratic governments dependent on popular vote for their legitimacy and continuance and being the policymakers for their economies might be tempted to intervene in the utilities market for political reasons, for instance, to keep prices low, or mandate extension of service to unremunerative segments of the population. But how to protect a private utility from such expropriation by an all-powerful government? Entrusting regulation to 'independent' commissions signalled 'credible commitments' on the part of the government to ensure fair return to investors making lumpy investments in network infrastructure to supply utilities like electricity or water. Thus, the USA, where government ownership of any business is frowned upon, had set up 'independent commissions' as early as the late eighteenth century and its contours, remit and role has evolved over time.[3]

The UK embraced independent regulation in the neoliberal era when it unbundled its vertically integrated utilities and privatized them under the stewardship of Margaret Thatcher in the 1980s. Privatization called for 'credible commitment' by the British government to ensure that the new investors in utility businesses were fairly compensated and that there were no 'hold-up' problems. The EU members also set up independent regulators under the common EU mandate for integration of network utility services across Europe. The World Bank and IMF took the lead in designing, implementing and monitoring of structural reforms in network utilities worldwide—in Latin America, former colonies of Africa and Asia, privatizing ownership where feasible and setting up country-level independent energy regulators.

[2] In order to ensure a fair rate of return to the investor, the regulator might want to commit to a reasonable rate of return ex-ante, whereas governments might create a hold-up by reneging on such commitments ex-post.

[3] Municipal utilities were the exception. While there are over 1600 municipally owned utilities not subject to regulation in the USA, these utilities serve a small segment of the population.

2.4.1.1 Credible Commitments

The theory of 'credible commitment' was first mooted by Douglass C. North in an article entitled 'Institutions and Credible Commitments' published in 1990 (North 1990). Central to North's argument is the inviolability of property rights, which are 'historical stumbling blocks' to realizing the potential gains from trade. He concludes that "the objective of restructuring must be the creation of an adaptively efficient economy, that is, one that over time, will provide an institutional framework for a wide menu of alternative choices for organizational innovation and also wipe out failures". Credible commitment would require the institution to get prices right, every time, over time. Majone, however, believes that the commitment problem is central to the study of transaction cost politics, not only because it takes care of ex-post opportunism, but of the very nature of the political process (Majone 1994).

When it comes to utility regulation, policymakers suffer from the inability to credibly commit to future policies (Coase 1972; Kydland and Prescott 1977) and from the existence of information that is privately known to the agents subject to their policies (Baron and Myerson 1982). Time inconsistency occurs when a government's optimal long-run policy differs from its optimal short-run policy, so that the government in the short run has an incentive to renege on its long-term commitments. Without a binding contract holding it to the long-run policy, the government will use its discretion to switch to what now appears to be a better policy.

The problem is that if economic actors anticipate such a policy change, they will behave in ways that prevent policymakers achieving best results. These two obstacles, 'time inconsistency' and 'asymmetric information', make it difficult, if not impossible, for government regulation to achieve first-best policies. The time-inconsistency problem in this context stems from the possibility of regulatory hold-up in rate-of-return regulation. The regulator would like to commit to a fair return on irreversible investments ex-ante. Once the investments are sunk, the regulator is tempted to adjudicate a lower return than promised, thereby expropriating sunk investments (Baron and Besanko 1987; Lewis and Sappington 1991; Gilbert and Newbery 1994). The utility realizes this dynamic, resulting in under-investment leading to an ageing infrastructure prone to too many power outages.

Such difficulty could be mitigated by delegating regulation to 'independent' actors who, nevertheless, govern through a set of predetermined rules (Levy and Spiller 1994). The task of the independent regulator, when defined narrowly, is to limit the scope for arbitrary administrative action, signaling a favourable investment environment to expand and upgrade infrastructure. Apart from statutory regulatory mandates and the regulations framed under them, rules would also include contractual arrangements between stakeholders.

While government regulation is hostage to discretionary decision-making, independent regulators have no such luxury. This objective of this book is to examine how regulators, government and the civil judiciary view their role in credibly committing to long-term contracts.

2.4.1.2 Rule by Experts

The other attribute of independent regulation is that it is conducted by experts and, as such, is also useful in dealing with information asymmetry. The assumption here is that sector experts might be able to unearth and marshal relevant information through insightful and persistent efforts which the government or bureaucracy is not in a position to do. Finally, independent regulator also gives a legitimate escape route for a government forced to take unpopular decisions (Thatcher and Sweet 2002). Thus, the independence of the regulator is justified by the nature of the functions they are expected to perform, the expertise they bring to the job and, finally, their insulation from political interference.

Independent regulation straddles the divide between public and private interest theories of regulation and is best defined as 'stakeholder regulation' (Prosser 2005) although that is contingent upon the nature of regulatory governance, whether it is based on robust procedures, etc. Yet another theory puts the regulator at the centre of conflicting interests of a spectrum of stakeholders—government, the regulated utilities and the consumer. It is the contention of this book that stakeholder regulation is a misnomer because of the mismatch in the relative strengths and organization of the different stakeholders. For instance, the consumers who eventually pay for utility services, are not organized, have no wherewithal to counter the acute information asymmetry and do not have the resources to hire experts to argue their case. They are seldom represented in regulatory hearings, if at all. Therefore, the regulator often errs on the side of the consumers and, at times, is even mandated to do so by statute. The centrality of the consumer in regulatory governance is self-evident. On the one hand, FERC orders that heating not be turned off during peak winter even when consumers have failed to pay their bills. Developing country governments, on the other hand, have devised their own mechanisms to bypass independent regulators to promote consumer interests when the latter fails to do so.

References

Alchian, Armen A. 1961. *Some Economics of Property*, No. P-2316. Rand Corporation.

Baron, David P., and David Besanko. 1987. Monitoring, Moral Hazard, Asymmetric Information, and Risk Sharing in Procurement Contracting. *The RAND Journal of Economics* 18 (4): 509–532. https://doi.org/10.2307/2555639. Accessed June 9, 2022.

Baron, David P., and Roger B. Myerson. 1982. Regulating a Monopolist with Unknown Costs. *Econometrica* 50 (4): 911–930.

Bonbright, James C. 1961. *Principle of Public Utility Rates, Public Utility Reports.* Arlington, Virginia.

Breyer, Stephen. 1984. *Regulation and Its Reform.* Harvard University Press.

Coase, Ronald H. 1960. The Problem of Social Cost. *Journal of Law and Economics* 3: 1–44.

Coase, Ronald H. 1972. Durability and Monopoly. *The Journal of Law & Economics* 15 (1): 143–149. The University of Chicago Press.

Crandall, Robert W. 1983. *Controlling Industrial Pollution: The Economics and Politics of Clean Air.*

Demsetz, Harold. 1967. Towards a theory of property rights. *The American Economic Review*, vol. 57, No. 2, Papers and Proceedings of the Seventy-ninth Annual Meeting of the American Economic Association (May, 1967), 347–359.

Dixit, Avinash K. 1996. *The Making of Economic Policy. A Transaction-Cost Perspective*. Cambridge, MA: The MIT Press.

Feintuck, Mike. 2010. *The Public Interest in Regulation*. Oxford University Press.

Gilbert, Richard J., and David M. Newbery. 1994. The Dynamic Efficiency of Regulatory Constitutions. *The RAND Journal of Economics* 25 (4): 538–554.

Groenewegen, John, and Jack J. Vromen. 1996. A Case for Theoretical Pluralism. In *Transaction Cost Economics and Beyond*, 365–380. Dordrecht: Springer.

Joskow, Paul. 2007. Regulation of Natural Monopoly. *Chapter 16 in Handbook of Law and Economics*, vol. 2, 1227–1348. Elsevier.

Kydland, Finn E., and Edward C. Prescott. 1977. Rules Rather than Discretion: The Inconsistency of Optimal Plans. *Journal of Political Economy* 85 (3): 473–492.

Laffont, Jean-Jacques, and Jean Tirole. 1993. *A Theory of Incentives in Procurement and Regulation*. Cambridge, MA: The MIT Press.

Levy Brian, and Pablo Spiller. 1994. The Institutional Foundations of Regulatory Commitment: A Comparative Analysis of Telecommunications Regulation. *The Journal of Law, Economics, and Organization* 10 (2): 201–46.

Lewis, Tracy R., and David E.M. Sappington. 1991. Technological Change and the Boundaries of the Firm. *The American Economic Review* 81 (4): 887–900.

Majone, Giandomenico. 1994. The Rise of the Regulatory State in Europe. *West European Politics* 17 (3): 77–101. https://doi.org/10.1080/01402389408425031.

McKenzie Richard B., and Hugh H. Macaulay. 1980. A Bureaucratic Theory of Regulation Public Choice, vol. 35, No. 3 (1980), 297–313 (17 pages). Springer.

North, D. 1990. *Institutions, Institutional Change and Economic Performance*. Cambridge: Cambridge University Press.

Peltzman, Sam. 1976. Towards a More General Theory of Regulation. *Journal of Law and Economics* 19: 211–240.

Posner, Richard A. 1971. Taxation by Regulation Bell. *Journal of Economics and Management Science* 2 (1): 22–50.

Posner, Richard A. 1974. Theories of Economic Regulation. *Bell Journal of Economics and Management Science* 5: 335–358.

Prosser Tony. 2005. Regulatory Contracts and Stakeholder Regulation. *Annals of Public and Cooperative Economics* 76 (1): 35–57.

Stigler, George J. 1971. The Theory of Economic Regulation. *Bell Journal of Economics and Management Science* 2: 3–21.

Thatcher, Mark, and Alec Stone Sweet. 2022. Theory and Practice of Delegation to Non-Majoritarian Institutions West European Politics, 25 (1).

Williamson, Oliver. 1981. The Economics of Organization: The Transaction Cost Approach. *American Journal of Sociology* 87 (3): 548–577.

Chapter 3
Sanctity of Contracts and the Hold-Up Problem

This chapter analyses the tortuous legal dispute relating to power purchase contracts signed by power utilities with three large power generation projects in coastal Gujarat, a Western Indian state. In the process, it attempts to answer the following questions: What are the circumstances that lead to demands for reopening of Power Purchase Agreements? Is the assumption that governments/government-owned entities always act in the public interest (in this case, consumer interest) borne out by evidence on the ground? Do regulatory judgements address consumer interests as mandated by the statute? What were the areas of convergence and divergence in the perspectives of various stakeholders on the issue of sanctity of contracts?

Regulatory theory makes a distinction between time-inconsistent and time-consistent contracts. In a seminal article published in 1977, Kydland and Prescott argue that a time-inconsistent policy may make the public happy in the short run, but will ultimately fail to produce long-run policy goals (Kydland and Prescott 1977). To illustrate, they point out how victims of flooding caused by illegal construction on floodplains, when bailed out during floods out of compassion, send out the wrong signals—that it is okay to build on floodplains, one would eventually be bailed out. If somehow, threats of no relief during flooding could be made credible, people would think twice before constructing on floodplains (ibid.).

Sending out such messages would require a framework and credible rules. Economists categorize policymaking frameworks as discretionary or rule based. Milton Friedman argued, in the context of monetary policy, that real-world policymakers have imperfect information and imperfect tools (Friedman 1968). In the event, he argued, there should be a prespecified plan to deal with consequences arising out of imperfect information.

The power sector too is hostage to imperfect information, and long-term contracts are particularly vulnerable to information asymmetry. In this chapter, we discuss 'time-inconsistent' contracts, wherein short-term considerations blind the contracting parties to potential long-term adverse developments which could eventually lead to demand for renegotiation of the contract by either party. Uncertainties are inherent in long-gestation, long-life infrastructure projects like power plants,

S. Mahalingam and K. Mahalingam, *Independent Energy Regulation in a Developing Economy*, https://doi.org/10.1007/978-981-99-4398-2_3

especially those using imported fuels. Contracting for such projects requires vision, diligence, appropriate tools and a degree of flexibility. Besides, as Laffont suggests, the institutional framework in many developing countries makes it impossible to rely on contracts. Contract renegotiation is rampant, especially in developing countries. Guasch, Laffont and Straub investigated why, in Latin America between 1985 and 2000, more than 40% of concessions (excluding the telecom sector) were renegotiated, a majority, at the request of governments. They concluded that fear of future renegotiation is a serious impediment to attracting private investments (Guasch et al. 2002).

In this section, we examine three specific demands for contract renegotiation in India's electricity sector. (It might be instructive to note that in the Indian case, the demand for renegotiation, thus far, has come not from the government as was the case in Latin America, but from the contracting parties in the private sector.) In this and the following chapters, we track multiple case studies where sanctity of contracts has been assailed for one reason or another, in order to find answers to the following questions:

(i) What are the circumstances that lead to demands for reopening of Power Purchase Agreements?
(ii) Is the assumption that governments/government-owned entities always act in the public interest (in this case, consumer interest) borne out by evidence on the ground?
(iii) In the new paradigm of deregulation, independent regulators are specifically tasked to protect consumer interests. Do regulatory judgements reflect this mandate?
(iv) What were the areas of convergence and divergence in the perspectives of various stakeholders on the issue of sanctity of contracts?

In 1992, when India liberalized its power generation sector and had opened it to potential foreign investors, it had adopted the MoU route wherein developers like Enron Corporation entered the Indian generation segment on the back of cost-plus Power Purchase Agreements with the state utilities, their payment security protected by sovereign guarantees. The two-part tariff policy announced around the same time ensured that power plant developers were compensated for fixed costs of setting up a power plant through a fixed tariff and fuel costs through a variable charge. Enron was the first IPP to enter the Indian market through this route. However, owing to its many contradictions, the Enron contract unravelled. Abhay Mehta provides an excellent analysis of the failure of the Enron project in his book Power Play (Mehta 2000). Few international investors ventured into India's power generation business thereafter, leaving it to public sector NTPC and other state government-owned generators to take up the slack.

With growing demand–supply mismatch, Government of India decided to invite private sector participation in generation through the competitive bidding route. In 1995, GoI had issued guidelines for competitive bidding for awarding these projects. These were revised in 2006 to align them with the provisions of Electricity Act 2003. Several IPPs have since been set up in various states through the competitive bidding route. Studies found that power plants awarded through ICB route have lower tariffs than the projects that came up through the MoU route (Prayas 2011).

According to Section 62 of Electricity Act 2003, the electricity regulator in each state would determine the tariffs of power plants set up under the MoU route for supply within that state as also the tariffs for transmission and distribution of power within the state boundaries. Similarly, CERC would determine the tariffs of power plants supplying more than one state as well as interstate transmission tariffs. However, the regulator's remit to determine power generation tariff in the case of competitively bid projects is circumscribed.

Section 63 of Electricity Act 2003 reads:

> Notwithstanding anything contained in Section 62, the Appropriate Commission shall adopt the tariff, if such tariff has been determined through transparent process of bidding in accordance with the guidelines issued by the Central Government.

Thus, in the case of tariffs discovered through competitive bidding, the regulators' remit is limited to ensuring compliance with government's bidding guidelines and approving the tariffs so discovered.

To meet the growing demand for electricity, in January 2005, GoI issued a new set of guidelines outlining the 'Bidding Process for Procurement of Power by Distribution Licensees', under Section 63 cited above (Annexure 2). This would be applicable to ultra-mega power projects (UMPPs), i.e. 4000 MW and above supercritical thermal power projects (both pit head and imported coal based). The bidders were given the option to assume responsibility for fuel linkage, size and location or assign that responsibility to the procurer discom and bid only fixed charges. By virtue of their large capacity, UMPPs were envisaged to supply to multiple states and, as such, would be treated as 'composite' projects to be regulated by CERC, the central regulator, not the respective state regulators. Power Finance Corporation Ltd. (PFC) was appointed as the Nodal Agency to facilitate the development of these projects. A special purpose vehicle would be created to coordinate and facilitate clearances such as water linkage, land allocation and environmental clearances. PFC would also arrange coal linkage where sought by the prospective investors and will facilitate Power Purchase Agreements (PPA) between the procurer states for all bidders.

UMPP bidding guidelines differ from the earlier 1995 guidelines in key respects. UMPPs are typically large-scale plants involving sizeable upfront capital costs and, once commissioned, could supply power continuously for several decades to multiple states. Acknowledging this, the bid design for such projects would have to accommodate the uncertainties associated with bidding for large, long-term projects. Accordingly, GoI guidelines are designed to address multiple scenarios and offer multiple options to bidders.

Broadly, bidding can be conducted under Case 1 or Case 2. Bids under Case 1 leave it to the power plant developer to decide and arrange the location, technology and fuel for the project. Case 2 guidelines address hydropower projects, load centre projects or other location-specific projects with specific fuel allocation such as captive mines or assumption of fuel supply responsibility by the procurer discom. Guidelines specify that tariffs will have to be usually two-part (fixed and variable) with exceptions only for medium term supply (1–7 years) for which single part tariff bid is an option. The guidelines also offer the developer, the option to bid for 'escalable' and/or 'non-escalable' tariffs or even partly escalable and partly firm tariffs, both for the fixed and variable part of the two-part tariffs design. Additionally, the guidelines contain an option for 'transfer price of fuel' if it has not been fixed by the government, government-approved mechanism or the fuel regulator, i.e. pass through all actual fuel costs without having to hark back to the regulator every time the fuel prices changed, by linking them to an index.

Section 4.11 of the guidelines states "Where applicable, the energy charges payable during the operation of the contract shall be related on the base energy charges specified in the bid with suitable provision for escalation. In case the bidder provides firm energy charge rates for each of the years of the contract term, the same shall be permitted in the tariffs". Thus, the option to quote fixed or pass-through fuel charges rested with the bidder. In cases where escalation is opted for, the bid document also states that the median escalation rate of the relevant fuel index in the international market for the last 30 years for coal and 15 years for gas/LNG shall be used for escalating the energy charge quoted by the bidder. While these provisions are considered at the bidding stage, the actual escalation through the life of the project is based on an index published by CERC every six months without prejudice to the winning bidder. Section 4.3 of the guidelines categorically states that all tariff will have to be bid in Indian Rupees only and any foreign exchange risk will have to be borne by the power developer for the fixed part component of tariff.

Yet, despite such clear bidding guidelines, the most egregious and protracted disputes brought before the regulator for adjudication relate to UMPP and other large projects straddling more than one state bid in accordance with GoI guidelines and awarded to the bidder quoting the lowest levelized tariff through the life of the project.

In this chapter, we examine three such projects all of which came up in coastal Gujarat—Coastal Gujarat Power Limited (CGPL) a UMPP promoted by Tatas and supplying to multiple states, a very large project of more than 4000 MW capacity that came up in different phases in the same location in coastal Gujarat and promoted by the Adani group, also supplying to multiple states and a third project of 1000 MW promoted by Essar group in Salaya, also a coastal location in Gujarat, but supplying only to Gujarat. All three were based on use of imported coal wholly or substantially which is why they were set up in a coastal location near port facilities. Of course, they would supply power to other states as well.

In 2006, Haryana Utilities represented through HPGCL initiated a competitive bidding for purchase on long-term basis, 2000 MW of electricity required for maintaining the supply to consumers in the State of Haryana. The participating bidders had an option to quote escalable or non-escalable tariff, or partly escalable and partly non-escalable tariff to cover their risk as per their own perception.

On 24.11.2007, Adani Power submitted its bid to HPGCL for supply of 1424 MW of power from its 4620 MW Mundra Power Plant (APMuL) to Haryana Utilities at a levelized tariff of Rs. 2.94/kWh from Units 7, 8 and 9 of Phase IV. Adani Power quoted non-escalable levelized/uniform capacity charges of Rs. 0.997 per kWh and quoted non-escalable levelized energy charges of Rs. 1.963 per kWh, aggregating to Rs. 2.940 per kWh for the entire life of the project. Adani Power did not opt for any escalation in the tariff for 25 years period either in the capacity charges or in the variable/energy charges. The bid of Adani Power was based on blend of domestic and imported coal in the ratio of 70:30. In support of the fuel linkage, Adani Power submitted copies of MoUs it had signed with M/s. Coal Orbis Trading GMBH and Kowa Company Limited of Japan, respectively. However, subsequently, this was superseded on 14.12.2009, by a coal supply agreement (CSA) was executed between Adani Global Pte Ltd. and PT Dua Samudera Perkasa for supply of 10 MMTPA of coal to meet all of APMuL's import requirements. APMuL informed Haryana Utilities that it would get imported coal from its own mines in Indonesia.

The Coal Linkage Committee tasked with allocation of domestic coal had decided that projects considered as coastal projects would have an import component of 30% for which the developer had to tie up sources directly, and Letter of Assurance would be issued for 70% of the recommended capacity only. For obtaining domestic coal, APMuL had made an application on 28.1.2008 to the Standing Linkage Committee (long term) (SLC-LT), Ministry of Coal, Government of India, for long-term coal linkage. SLC-LT authorized issuance of LOA for supply of 6.049 million tonnes/year by Coal India Limited from Mahanadi Coalfields, to meet the requirements of 1386 MW for Phase IV of the project (70% of installed capacity of 1980 MW) in accordance with the provisions of new coal distribution policy.

However, despite this agreement, APMuL was informed subsequently that supply of coal under CSA from domestic sources was not likely to exceed 80% of Annual Contracted Quantity (i.e. 80% of 70% of plant capacity, which is 56% only), and the balance should be sourced through import, a condition to which APMuL agreed without changing its original bid tariff. On 17.7.2008, Adani Power was declared the successful bidder and awarded the contract on the basis of firm (non-escalable) tariffs for both fixed and fuel cost for the entire life of the project, viz. 25 years.

APMuL had already signed an earlier PPA with GUVNL for supply of 1200 MW from its units 1 to 6 to Gujarat discoms from a power plant to be built in Chhattisgarh using domestic coal from Morga II in Korba, a block allotted to Gujarat Mineral Development Corporation (GMDC). However, ten days after signing the PPA, Adani decided to shift this project too to Mundra in Gujarat alongside the other Adani project that would supply to Haryana and other utilities from units 7 to 9. This proposal was accepted by GUVNL. Thus, Adani Power Mundra became a contiguous project of nine units of which five would be based on supercritical technology, all in the same location. Including the original project that was shifted from Chhattisgarh, Adani Power Mundra (APMuL) comprised a total of 4620 MW, although it was not categorized as UMPP (see Chap. 5).

3.1 CGPL

Around the same time in 2006, another 4000 MW UMPP called Coastal Gujarat Power Limited (CGPL) was planned by Power Finance Corporation, a Government of India company through a special purpose vehicle which would be handed over to the successful bidder. This project too would be located in Mundra in coastal Gujarat and would supply power to multiple states, namely Gujarat, Maharashtra, Rajasthan, Punjab and Haryana. Six bidders responded to the Request for Proposal (RfP) including Tata Power. After evaluation of all the bids, Tata Power was declared as the successful bidder, it having quoted the lowest levelized tariff of 2.26367/kWh. CGPL entered into two PPAs dated 2.2.2007 and 6.2.2007 for supply of 2 × 1000 MW power to Gujarat Urja Vikas Nigam Limited (GUVNL), the holding company of Gujarat discoms, from Phase I, II and III and another PPA dated 7.8.2008 with Uttar Haryana Bijli Vidyut Nigam Ltd. and Dakshin Haryana Bijli Vidyut Nigam Ltd. (Haryana Utilities) for supply of 1424 MW power from Phase IV of the generating station.[1]

Under Section 63 of Electricity Act 2003, tariffs of all composite projects straddling more than one state awarded on the basis of competitive bids will have to be formally approved by CERC, the central regulator. In 2007, CERC formally accepted the tariff quoted by Tata Power, having found it was the result of a transparent process of bidding in conformity with GoI guidelines.

CGPL was to be executed entirely on the basis of imported coal with an estimated coal requirement of approximately 12 MMTPA. According to CGPL, it had made arrangement for importing coal from Indonesia and had entered into CSA dated 31.10.2008 with IndoCoal Resources (Cayman) Limited for supply of 5.85 MMTPA (± 20%). Tata Power had made a strategic investment of USD 1.2 billion in Indonesia in February 2007 for acquiring 30% stake in PT Kaltim Prima Coal (KPC), PT Arutmin Indonesia and IndoCoal Resources (Cayman) Ltd. from PT Bumi.[2] Tata Power assigned these in favour of CGPL.

Once the successful bidder is awarded the contract, a Power Purchase Agreement is signed between the developer and the procurers. In Case 1 bids where no escalation is sought, the levelized tariff is sealed for the life of the project, viz. 25 years. Considering there were six final bidders for CGPL and eight for APMuL, both Adani and Tata had won their respective bids because of the low and stable (non-escalable over 25 years) tariffs they had quoted. In fact, Prayas Energy Group in its report dated March 2011 has expressed concerns over the viability of many projects that had been awarded on the basis of exceptionally low tariffs quoted presumably to win the bid (Prayas 2011).

Once a bidder wins the contract on the basis of non-escalable tariffs, there are only two exceptions to this condition, namely any 'Change in Law' (Article 12 of the PPA) whereby a legal action of a government body or a court imposes any cost (or results in benefit) and 'Force Majeure' (Article 13 of the PPA), which implies an unforeseen

[1] CERC order dated 21.2.2014.

[2] Dissent note by R. Jayaraman, Member CERC (Annexure 3).

event which prevents or unavoidably delays the performance of obligations under the contract. In such a situation, escalation/reduction in tariffs could be allowed.

12.1.1 'Change in Law' means the occurrence of any of the following events after the last date of bid submission resulting into any additional recurring/non-recurring expenditure by the SPD or any income to the SPD:

- the enactment, coming into effect, adoption, promulgation, amendment, modification or repeal (without re-enactment or consolidation) in India, of any law, including rules and regulations framed pursuant to such law;
- a change in the interpretation or application of any law by any Indian Governmental Instrumentality having the legal power to interpret or apply such law, or any Competent Court of Law;
- the imposition of a requirement for obtaining any Consents, Clearances and Permits which was not required earlier;
- a change in the terms and conditions prescribed for obtaining any Consents, Clearances and Permits or the inclusion of any new terms or conditions for obtaining such Consents, Clearances and Permits; except due to any default of the SPD;
- any statutory change in tax structure, i.e. change in rates of taxes, duties and cess, or introduction of any new tax made applicable for setting up of Solar Power Project and supply of power from the project by the SPD and has direct effect on the project, shall be treated as per the terms of this agreement; but shall not include
  (i) any change in taxes on corporate income or any withdrawn tax on income or dividends distributed to the shareholders of the SPD, or
  (ii) any change on account of regulatory measures by the Appropriate Commission.

12.2 Relief for Change in Law

12.2.1 The aggrieved Party shall be required to approach the Appropriate Commission for seeking approval of Change in Law.

12.2.2 The decision of the Appropriate Commission to acknowledge a Change in Law and the date from which it will become effective, provide relief for the same, shall be final and governing on both the Parties.

3.2 Force Majeure

Available Relief for a Force Majeure Event Subject to this Article 12:

No party shall be in breach of its obligations pursuant to this agreement to the extent that the performance of its obligations was prevented, hindered or delayed due to a Force Majeure Event.

Every party shall be entitled to claim relief in relation to a Force Majeure Event in regard to its obligations, including but not limited to those specified under Article 4.5.

For the avoidance of doubt, it is clarified that no tariff shall be paid by the procurers for the part of Contracted Capacity affected by a Natural Force Majeure Event affecting the seller, for the duration of such Natural Force Majeure Event. For the balance part of the Contracted Capacity, the procurer shall pay the tariff to the seller, provided during such period of Natural Force Majeure Event, balance part of the power station is declared to be available for scheduling and dispatch as per ABT for supply of power by the seller to the procurers.

Until 2010, coal mining companies in Indonesia were free to price the coal they mined after payment of royalties and taxes to the Indonesian government. We have no information on the rate at which Tata CGPL and Mundra Adani priced coal from their own mines when they bid energy charges for the power projects they won in Gujarat; nor are we privy to any information on the assumptions on which they made their respective bids. In a competitive bid, the final price (at which power will be supplied at the bus bar, in this instance) split into fixed and variable tariffs is the sole bidding parameter, not how it is arrived at. It is up to the bidder to factor in all the risks since he is presumed to have all the necessary information to arrive at the bid price, whereas the procurer utility has little information, acute information asymmetry being a perennial problem in infrastructure industries. It is precisely to overcome such information asymmetry problems that Harold Demsetz advocated a competitive bidding process to select the project developer (Demsetz 1968). It has been adopted in greenfield projects worldwide.

On 23.09.2010, the Ministry of Energy and Mineral Resources, Indonesia, notified a regulation # 17 of 2010 dealing with Coal Benchmark Export Price (Indonesian Regulation) to be effective from 01.09.2012. The notification mandated benchmarking export price of Indonesian coal to international coal prices indicated an index called HBA. All coal exporters from Indonesia were required to align their export prices to international benchmarks within a year of such notification (six months in the case of shorter duration contracts) failing which they would attract penalties and, eventually, cancellation of mining licenses. The increased price realization of coal sold at benchmark prices was allowed to be retained by the coal exporting company in Indonesia, after payment of proportionate royalty and taxes to the Indonesian government.

The promulgation of Indonesian Regulation resulted in escalation in the export price of Indonesian coal. CGPL claimed that while its tariff quoted by it was fixed for 25 years under the PPA, the company was forced to burn Indonesian coal at global benchmark prices, leading to losses. The company claimed that without adjustment of tariff, it would suffer a loss of Rs. 1873 crores per annum and Rs. 47,500 crores over a period of 25 years. APMuL which envisaged using 44% imported coal in its project claimed that coal supplies from its Indonesian mines pushed up fuel prices by 153% consequent to the Indonesian notification, rendering awry, the basis of its calculations in bidding for the Mundra project.

CGPL took up the matter with GUVNL, its lead procurer and the Ministry of Power, GoI vide letter dated 04.08.2011. Ministry of Power in its reply dated 30.09.2011 responded to CGPL's representation stating "... PPA is a legally binding document exclusively between the procurers and the developer. Therefore, any issue arising therein is to be settled within the provisions of the PPA by the contracting parties for which Gujarat being the lead procurer may take necessary action ...". CGPL also took up the matter with the procurers in a Joint Monitoring Meeting dated 06.02.2012 for suitable adjustment in tariff. The procurers sought some further details which CGPL furnished in its letter dated 06.03.2012. CGPL had also approached the Indonesian government vide its letter dated 16.02.2012 requesting to exempt the existing coal supply contracts from the purview of Indonesian Regulation, without any success.[3] APMuL also took it up with the procurers as well as the GoI, seeking compensatory tariffs to make up for the escalation in coal costs, but received a similar rejection from both. Left with no option, both CGPL and APMuL petitioned the regulator for redressal of their grievances.

APMuL had an additional issue in terms of reduced availability of domestic coal for its project which meant higher import of coal and, thus, further increase in its costs. Therefore, it claimed that the PPAs with the distribution companies in Haryana quoted a tariff of Rs. 2.94/kWh based on central government's policy of allocation of coal linkage for the power projects, the market conditions relating to the price and availability of fuel at the time of the bids. Adani stated that after submission of RFP to HPGCL, it made an application to Coal India Ltd. on 28.1.2008 for grant of coal linkage and was issued Letter of Assurance (LoA) for coal linkage equivalent to 70% of capacity although prior thereto the central government had proposed 100% allocation of coal to the power projects. Thus, 30% of the coal used in its plant was impacted by the Indonesian notification and had rendered all its calculations awry. APMuL filed a petition with CERC (155/MP/2012) seeking the following relief:

(a) to evolve a mechanism to restore the petitioner (APMuL) to the same economic condition prior to occurrence of subsequent events mentioned in respective Parts I and II hereinabove (Indonesian coal notification) by adjudicating the disputes between the petitioner (APMuL) and the discoms in relation to regulate including changing and/or revising the price/tariff under PPAs dated 7.8.2008 with UHBVNL and DHBVNL and 2.2.2007 with GUVNL;
(b) in the alternative, to declare that the applicant is discharged from the performance of the PPAs on account of frustration of the PPAs due to subsequent events in respective Parts I and II;
(c) to declare that the revised tariff shall be applicable from the Scheduled Commercial Operation Date (SCoD) of the PPAs;
(d) that during the pendency of the case, the commission may direct the respondent(s) to procure power on the cost-plus basis, alternatively, may suspend the operation of the PPAs till the final disposal of the application.

[3] Extracted from CERC order dated 21.2.2014 (155/MP/2012).

CERC heard the parties and issued an order dated 2.4.2013 (ibid.). With regard to the domestic coal, the commission came to the conclusion that availability of coal from CIL was posing a challenge as CIL had expressed its inability to supply the desired quantum of coal causing difficulties for the power plants to even meet their minimum coal requirement equivalent to normative availability. However, the regulator held that Indonesian Regulations did not constitute ‘Force Majeure’ or ‘Change in Law’ in terms of the PPAs with GUVNL and Haryana Utilities. Non-availability of full coal linkage from CIL or its subsidiaries cannot be considered as a Force Majeure Event. Further, since the petitioner applied for linkage of domestic coal to Coal India Ltd. on 28.1.2008 after the petitioner was awarded LoI by GUVNL and Haryana Utilities, it could not be said that the bids were premised on domestic coal linkage, and hence, the change in policy of GoI/CIL could not be considered ‘Change in Law’.

However, the commission in discharge of its statutory functions under Section 79 of the 2003 Act which empowers it to intervene in the interest of the consumers, investors and the power sector, coupled with provisions of the National Electricity Policy and tariff policy mandating the regulator to ensure financial turnaround of the power sector and supply of electricity to the consumers at competitive prices, CERC held that it was statutorily obliged to work out a solution to help the petitioner to get over the unprecedented situation arising out of the promulgation and implementation of the Indonesian Regulations.

The commission decided to grant of relief in the form of compensatory tariff over and above the tariff agreed in the PPAs which would be admissible for a limited period till the event which occasioned such compensation continued to exist and ordered it should also be subject to periodic review by the parties to the PPAs.

Therefore, the regulator constituted an expert committee to look into the claims for compensatory tariffs for APMuL, over and above the PPA tariff. The expert committee would comprise, apart from representatives of the procurer utilities and the litigant, also two bankers who had lent to the project as well as independent experts and consultants.

After considering the suggestions and objections of the parties on the recommendations of the expert committee and the submissions made during the hearing, the commission issued the order dated 21.2.2014 quantifying the compensatory tariff to be paid by the procurers to the petitioner along with the mechanism for its recovery from the procurers.

However, one member of CERC dissented with the majority order in the APMuL petition and recorded his dissenting order separately wherein he maintained that from the RfP stage, it was clear that the bidder had the full freedom to quote either firm capacity charge or a combination of escalable and non-escalable capacity charges. Similarly, the bidder had the full freedom to quote firm energy charge for each of the 25 contract years or a combination of escalable and non-escalable energy charges. That being the case, the dissenting member found no reason to award compensatory tariffs over and above what was quoted by APMuL in the bid.

‘In my view, the present case primarily involves adjudication of disputes raised by the petitioner and is outside the scope of regulatory power. The regulatory power is a general power vested in the Commission which can be exercised while formulating regulatory policies. The regulatory power cannot be invoked for settlement of individual disputes arising out of commercial relations between the parties, though power of regulation is considered to be expansive and vast. None of the authorities relied upon by learned counsel for the petitioner involves exercise of regulatory power to upset the agreed commercial arrangements between the parties. The exercise of regulatory power amounts to invasion on the exercise of free will by the parties … There is another aspect which merits attention before invoking regulatory power. The decision in the present case will be the precedent to be followed in future. The exercise of regulatory power in such cases will have the cascading effect. The relief to the petitioner in any form in the present set of circumstances will impinge upon the avowed object of the law mandating protection of the consumer interest.

Therefore, I am opposed to exercise of regulatory power under clause (b) of sub-section (1) of Section 79 of the Electricity Act to redress the petitioner's grievances which are of the nature of self-inflicted wounds. … The petitioner should have built in the escalation factor and the risk associated with sourcing coal from foreign countries to insulate it from any future adverse development. The petitioner, by quoting non-escalable energy charges, has assumed the commercial risks and to corner the award of contract and in my view, renegotiation should not be allowed as it would give an opportunity to the petitioner to pass on the risks he assumed to the consumers and defeat the purpose of section 63 of the Act … In my view, the comparative higher price of electricity discovered through the competitive bidding process at present cannot be the ground for reopening the tariff discovered more than five years ago … The petitioner may have resorted to aggressive and predatory bidding to win the bids by edging out the other competitors for which the petitioner is accountable and the consumers of Gujarat and Haryana should not be made to pay for the miscalculation/mistake of the petitioner, if any, and to ensure profitability of the petitioner irrespective of assumption of commercial risks …'.

'The bid documents provided for opportunity to quote firm, partly variable and fully variable (variable according to the indices notified by CERC from time to time) for fixed as well as fuel charges. It is also seen that many tenderers have quoted variable prices whereas the petitioner has quoted firm prices for twenty years fully knowing the fluctuating market conditions with regard to price of coal as well as variations in Rupee exchange as bids were only in rupees. It is obvious that the petitioner has built up adequate provisions in the rates to cover the variations. It is also obvious that the petitioner is using the notifications of Govt. of Indonesia as an opportunity to cover some of its commercial risks or to improve his margins further … Permitting negotiation at this stage will not only render the bidding process redundant but will also open up a potent legal issue affecting the rights of the bidders who have been edged out even after quoting tariff lower than what is sought to be determined through renegotiation …' (Dissent Note in Annexure 3).

But the majority order prevailed and the expert committee was tasked to work out the quantum and modalities of supplementary tariffs to APMuL on 2.4.2013. Meanwhile, in 2012 itself, CGPL had also approached CERC for compensatory tariffs to make up for the difference in fuel prices on account of Indonesian notification (159/

MP/2012). All the procurer discoms were made respondents to the petition. In its petition, CGPL claimed that the Indonesian notification constituted an event of Force Majeure under Article 12 of the PPA and also a Change in Law under Article 13 of the PPA. It prayed that the regulator

(a) Establish an appropriate mechanism to offset in tariff the adverse impact of (i) the unforeseen, uncontrollable and unprecedented escalation in the imported coal price and (ii) the Change in Law by Government of Indonesia.
(b) Evolve a methodology for future fuel price pass to secure the project to a viable economic condition while building suitable safeguards to pass to procurers benefit of any reduction in imported coal price.
(c) Pass any other order that this commission may deem fit in the facts and circumstances of the present case.

In CGPL case, CERC made the following order on 21.2.2014[4]:

> … we have come to the conclusion that the promulgation of Indonesian Regulations which required the sale price of coal in Indonesia to be aligned with the international benchmark price has, prima facie, altered the premise on which the energy charges were quoted by the petitioner in its bid. No doubt, the petitioner had taken huge risk by quoting 55% of the energy charges under non-escalable head as a result of which the benefits of escalation index are not available to the petitioner. Though the petitioner had quoted non-escalable energy charges to keep the bid price low, it was however factored on the basis of the then prevailing coal price for import from Indonesia … Moreover, quotation of low bid price was in the interest of the consumers as the power would be available at the levelized tariff of '2.26367/kWh to the respondents. The petitioner would have continued to supply power at this price, had the Indonesian Regulations not made it mandatory for sale of coal from Indonesia at international bench-mark prices. Therefore, the competitive advantage of hedging in coal prices that the petitioner was enjoying by acquiring mining rights in Indonesia or by entering into long term contract with the coal suppliers in Indonesia appears to have been fundamentally altered/ wiped out, after the coal sales are required to be aligned with international benchmark prices of coal. It is pertinent to note that the coal price in the international market is fluctuating. Therefore, the exact impact of the Indonesian Regulations will vary from time to time. We are also aware that other sources of imported coal are presently costlier than the Indonesian coal and it would not serve any purpose to say that the petitioner has got other viable options to source imported coal.
>
> the respondents shall be required to invite fresh bids to meet their requirement of power and till the selected project or projects are operationalised, the consumers will be deprived of power. Moreover, the ruling tariff for the new projects are in the range of '3.50–'7.00/kWh which the consumers of Mundra UMPP shall also be required to pay. Thus at the macro level, it will be a serious setback for the electricity sector and will adversely affect the investment for the sector and at the micro level, it will affect the continued and reliable supply of power to the consumers.

As to the relief sought, CERC concluded that the case did not fall within the purview of either Force Majeure or Change in Law. The commission strongly disapproved of the renegotiation of tariff as discovered through the competitive bidding and emphasized that the sanctity of the PPAs and the tariff agreed therein should be maintained. Yet, the commission decided to grant of relief in the form of compensatory tariff over and above the tariff agreed in the PPAs.

[4] https://cercind.gov.in/2014/orders/SO159.pdf accessed on 16th May 2023.

> we are not inclined to favour any re-negotiation of the tariff discovered through the process of competitive bidding as in our view, the sanctity of the bids should be maintained. The parties should not renegotiate the tariff discovered through the competitive bidding as that will bring uncertainty to the power sector and is prone to misuse. In our view, the parties should confer to find out a practicable solution and agree for compensation package to deal with the impact of subsequent event while maintaining the sanctity of the PPAs and the tariff agreed therein. In other words, the compensation package agreed should be over and above the tariff agreed in the PPAs and should be admissible for a limited period till the event which occasioned such compensation exist and should also be subject to periodic review by the parties to the PPAs. (ibid.)

CERC justified its order stating, "*The Electricity Act, 2003 vests in the Commission the responsibility to balance the interest of the consumers with the interest of the project developers while regulating the tariff of the generating companies and transmission licensees. Financial viability of the generating stations is an important consideration to enable them to continue to supply power to the consumers. The present case is one of the first of its kind where the tariff was determined through competitive bidding under Section 63 of the Act. The petitioner had quoted the bids on certain assumptions and those assumptions have been negated on account of the unexpected rise in coal price in international market coupled with the promulgation of Indonesian Regulations, required all long-term contracts to be adjusted to the international benchmark price. In our view, under the peculiarity of the facts of the present case and also keeping in view the interest of both project developer and consumers, we consider it appropriate to direct the parties to set down to a consultative process to find out an acceptable solution in the form of compensatory tariff over and above the tariff decided under the PPA to mitigate the hardship arising out of the need to import coal at benchmark price on account of Indonesian Regulations*" (ibid.).

Accordingly, CERC directed both CGPL and the discoms to constitute a committee consisting of the representatives of the state governments, distribution companies of the procurer states, the nominee of the Chairman of Tata Power Limited, an independent financial analyst of repute and an eminent banker dealing and conversant with infrastructure sector.

"The Committee shall go into the impact of the price escalation of the Indonesian coal on the project viability and obtain all the actual data required with due authentication from independent auditors to ascertain the cost of import of coal from Indonesia and suggest a package for compensatory tariff which can be allowed to the Petitioner over and above the tariff in the PPAs. The Committee shall keep in view inter-alia the following considerations while working out and recommending the compensatory tariff applicable upto a certain period:

(a) The net profit less Govt. taxes and cess etc. earned by the petitioner's company from the coal mines in Indonesia on account of the bench mark price due to Indonesian Regulation corresponding to the quantity of the coal being supplied to the Mundra UMPP should be factored in full to pass on the same to the beneficiaries in the compensatory tariff.

(b) The possibility of sharing the revenue due to sale of power beyond the target availability of Mundra UMPP to the third parties may be explored" (ibid.).

Once again, this order of CERC was signed by the chairman and three other members, while the fourth member, the same one who had issued a dissenting order in the similar petition filed by APMuL, appended a dissenting note. He agreed with the majority order that the case did not fall under either Change in Law or Force Majeure, but he disagreed with their conclusion that the developer had to be compensated for the hardship caused by increase in the price of coal.

The dissenting member noted that the CSA of Tata was already aligned to international price of coal, and that the Indonesian notification was not responsible for pushing up the price of coal to be used in CGPL, but the fact that global coal prices had risen. He therefore, observed that

> the loss in the case of escalable fuel price would get covered in future as the indices lag in movement compared to actuals and in a cycle they get adjusted. The petitioner being in business for a pretty long time is expected to factor in the possible market variation, while quoting for a period of more than 25 years. It can be concluded that a bidder who quoted firm price of power for a period of 25 years has safely assumed his own perception of the market variations with regard to fuel price and has suitably provided for the same in the tender quoting firm prices … The increase in price of coal directly benefits the Indonesian company which benefits are passed on to Tata Power Limited in the shape of return for the investment and thus the Tata Power Limited, as a whole may be the ultimate beneficiary of the Indonesian Regulations.

As for the majority decision to award relief under Section 79 of Electricity Act 2003, the dissenting member held

> The regulatory power cannot be invoked for settlement of individual disputes arising out of commercial relations between the parties, though power of regulation is considered to be expansive and vast. None of the authorities relied upon by learned counsel for the petitioner involves exercise of regulatory power to upset the agreed commercial arrangements between the parties. The exercise of regulatory power amounts to invasion on the exercise of free will by the parties.

Therefore, the dissenting member opined that the petition lacked merit and deserved to be dismissed (dissent note in Annexure 3).

The same committee appointed to give relief to APMuL was tasked to compute the quantum of relief in the form of compensatory tariff and the manner in which it would be recovered from the procurers. It comprised, apart from representatives of the power companies as well as discoms, two bankers, both financiers to the projects and an independent expert.

Curiously, it was CGPL which appealed to APTEL challenging CERC's rejection of 'Force Majeure' and 'Change in Law' even though the latter had ordered compensatory tariff under Section 79 of EA 2003. Even before the discoms could either implement or challenge CERC's order, Adani issued a notice for termination of PPA to GUUVNL. Consumer organizations—Prayas Energy Group and Consumer Watchdog—moved APTEL against the order of the regulator. Also aggrieved by CERC order were Haryana and Rajasthan Utilities with PPAs with UMPPs. They too appealed to APTEL against CERC order. APTEL clubbed all the petitions including one each of GMR Kamalganga and Sasan Power since the issues raised were similar and gave a consolidated order.[5]

The questions before APTEL, framed by APTEL itself as per its order, are as follows:

1. Whether the Central Commission, de-hors the provisions of the PPAs, has the regulatory powers to vary or modify the tariff or otherwise grant compensatory tariff to the generating companies in the case of a tariff determined under a tariff-based competitive bid process as per Section 63 of Electricity Act 2003?
2. Whether the Appropriate Commission, independent of Force Majeure and Change in Law provisions of PPAs, has the power to vary or modify the tariff or otherwise grant compensatory tariff to the generating companies in pursuance of its powers under Sections 61, 63 and 79 of the Act and/or Clause 4.7 and 5.17 of the Competitive Bidding Guidelines issued by the central government and/or Article 17.3 of the PPA and/or under the adjudicatory powers as per Section 79(1)(f) of Electricity Act 2003?
3. Whether, in the facts and circumstances of the case, the Central Commission having held that Force Majeure and Change in Law provisions of the PPAs have no application, is right in granting compensatory tariff under any other powers?
4. Whether in the facts and circumstances of the case, the Central Commission is right in construing the order dated 2.4.2013 in case of Adani Power and order dated 15.4.2013 in the case of CGPL as a decision of the commission to grant compensatory tariff not being limited to a conciliatory process to explore an amicable agreed solution which would exhaust if no consensus emerges?
5. Whether, in the facts and circumstances of the case, the Central Commission is right in giving effect to the payment of compensatory tariff retrospectively from the respective Scheduled COD of the generating units instead of considering the same prospectively from order dated 21.2.2014?
6. Whether, in the facts and circumstances of the case, the Central Commission is right in giving effect to the payment of compensatory tariff retrospectively from the respective Scheduled COD of the generating units instead of considering the same prospectively from order dated 21.2.2014?
7. Whether the Change in Law provided under Article 13 of the PPA or under Clause 4.7 of the said Guidelines issued by the central government as per Section 63 of the said Act should be construed to include laws other than Indian laws?
8. Whether in the facts and circumstances of the present case, the increase in price of coal on account of change in National Coal Distribution Policy linked to reduced availability of domestic coal and/or promulgation of Indonesian Regulation constitute an event of Change in Law?

[5] APTEL order dated 7 April 2016 on APPEAL NO. 100 OF 2013 and I.A. NO. 116 OF 2013; APPEAL NO. 98 OF 2014 and I.A. NO. 343 and 402 OF 2014.

9. Whether in the facts and circumstances of the case, the increase in price of coal on account of the intervention by the Indonesian Regulations as also the non-availability/short supply of domestic coal in case of Adani Power constitute a Force Majeure Event in terms of the PPA?
10. Whether the bid for generation and sale of electricity by Adani Power to GUVNL was premised on the availability of coal from GMDC and to what effect?
11. Whether the bid for generation and sale of electricity by Adani Power to Haryana Utilities for generation and sale was affected by non-availability of coal from Mahanadi Coalfields Limited and if so to what extent?
12. Whether the CGPL had Fuel Supply Agreements for procurement of coal for Mundra project at a price less than market price and if so to what was affected by non-availability of coal from Mahanadi Coalfields Limited and if so to what extent?
13. Whether the Central Commission has properly taken into consideration the mitigating factors in favour of the consumers and to the proper extent, namely, reduction in the return on equity, Sharing of Mining Profits in Indonesia, sale of quantum in excess of the target availability, residual value of the generating units at the end of 25 years, etc.? (ibid.).

Energy Watchdog argued that the generators were trying to convert determination of tariff through competitive bidding process under Section 63 of the said Act into determination of cost-plus tariff under Section 62 of the said Act, which must not be allowed. By quoting non-escalable costs, they became the lowest bidders ousting the other bidders. Now to upset the arrangement was unknown to law and was violative of Article 14 of the Constitution of India, it argued.

The consumer organizations also argued that the Central Commission should not have permitted Mr. Deepak Parekh, Chairman of HDFC, and Ms. Arundhati Bhattacharya, MD and CEO of SBI Capital Markets Ltd. to be members of the committee because they were the lenders of the generators. If Adani Power and CGPL became bankrupt on account of soaring coal costs, the banks would not have been able to recover their loans. Thus, the bankers in the committee had an interest in ensuring that the power companies got compensatory tariffs to make up for the impact of Indonesian notification. Besides, the committee's proceedings did not have provision for consumer participation.

CERC, one of the respondents in the dispute argued before APTEL that while the tariff of a project is determined through a Section 63 route, the Regulatory Commission has a limited role at the stage of determination, and it is required to adopt the tariff, during the lifespan of the project, it is still empowered to exercise regulatory powers as prescribed under Section 79(1)(b) of the said Act. It is not merely a dispute resolution body. It has been entrusted with multiple roles. Section 79(1)(b) provides the Central Commission with the power and also the duty to regulate all tariffs and the said power cannot be extinguished on the sole ground that the manner of determination of tariff is different (CERC submission).

CERC argued therefore, that in the emergent situation, to secure the viability of the generators' project, it was within its power to regulate. CERC further submitted that even the Supreme Court had held that Central Commission is required to adopt a pragmatic approach to ensure the viability of units when they are faced with an existential threat (CERC). CERC also cited the Kelkar Committee Report: "developing a mechanism for dealing with uncertainties inherent in long-time contracts. It must be kept in mind that given market and technological uncertainties, both public and private managers of long-term PPPs take decisions based on incomplete information. A decision that looks problematic ex-post cannot automatically be deemed to be mala fide".

APTEL considered the submissions and arguments of all litigants and defendants. A full bench order of APTEL running into 486 pages ruled thus on 7.4.2016:

Tariff discovered through bidding process cannot be tinkered with by resorting to Section 62(4) and Section 64(6) of the said Act. It is not possible to hold that the Change in Law provision contained in Clause 4.7 of the said Guidelines provide an independent right being a stand-alone provision, and the generators can claim that the Change in Law under the said clause has a wider implication than Article 13 of the PPA which refers to Change in Law so as to get a relief de-hors the PPA. Similarly, Clause 5.17 of the said Guidelines cannot be construed as providing a separately enforceable dispute resolution (arbitration) provision de-hors Article 17.3 of the PPA. The generators cannot get any relief by invoking Clause 4.7 or 5.17 of the said Guidelines de-hors the provisions of the PPA.

"In our view, the parties should confer to find out a practicable solution and agree for compensation package to deal with the impact of subsequent event while maintaining the sanctity of the PPA and the tariff agreed therein. In other words, the compensation package agreed should be over and above the tariff agreed in the PPA and should be admissible for a limited period till the event which occasioned such compensation exist and should also be subject to periodic review by the parties to the PPA".

The commission, going beyond the boundaries of the contract, imposed the entire burden of fuel price variation on to the consumers, directly or indirectly and allowed compensatory tariff to Tata Power and Adani Power, to the tune of Rs. 2300 crore and Rs. 3600 crore, respectively, till March 2016.

Following this order, regulatory commissions of several other states such as Maharashtra, Uttar Pradesh and Rajasthan adopted the same approach of modifying competitively set tariffs by granting compensatory tariffs. According to The Economic Times newspaper, the total amount of compensatory tariff granted till date would have amounted to around Rs. 11,000 crore.

Appellate Tribunal for Electricity (APTEL) rejected the use of regulatory power to grant relief to the projects. It held that changes in the domestic coal distribution policy and promulgation of the Indonesian Regulation couldn't be treated as 'Change in Law' under the PPA. It stated, the change in the price of imported coal would rather fall under the purview of 'Force Majeure'. By holding this, it directed the commission to determine the exact scope of relief to be granted to such projects. Subsequently, CERC through another order lessened the amount of tariff.

In its full bench order dated 7.4.2016, APTEL held.

"We hold that the increase in price of coal on account of the intervention by the Indonesian Regulation as also the non-availability/short supply of domestic coal in case of Adani Power constitute a Force Majeure Event in terms of the PPA'. APTEL further held that Adani's PPA with GUVNL was not solely premised on the availability of coal from GMDC. "Admittedly, Adani Power sourced coal from Indonesia to fulfill its contractual obligations. Accordingly, Issue No. 13 is answered in the negative. We also hold that the bid for generation and sale of electricity by Adani Power to Haryana Utilities was affected by non-availability of coal from Mahanadi Coalfields Limited. The shortfall in domestic coal was made good by Adani Power by importing Indonesian coal".

"We remand Petition No. 155/MP/2012 filed by Adani Power and Petition No. 159/MP/2012 filed by CGPL to the Central Commission and direct the Central Commission to assess the extent of impact of Force Majeure Event on the projects of Adani Power and CGPL and give them such relief as may be available to them under their respective PPAs and in the light of this judgment after hearing the parties".

On 16.08.2013, the expert committee set up by CERC submitted its report before the Central Commission. The report recommended a package of 'compensatory tariff' measures to be awarded to the companies and paid for by the discoms which, in turn, would be passed on to the consumers through higher electricity bills. These recommendations were accepted by the CERC in its order dated 21.2.2014. The CERC order reduced the fixed tariff by 20 paise/kWh but allowed the increased fuel cost to be passed on fully to the consumers in the respective discoms' jurisdictions. All the procurers except Punjab accepted the HPC report with some modifications. However, Punjab rejected the recommendations of the committee. A final order dated 6.12.2016 was issued by CERC, awarding compensatory tariff to the power generators.

APTEL's order was challenged in the apex court by a catena of stakeholders—consumer organizations like Energy Watchdog and Prayas Energy Group, Punjab, Rajasthan and Maharashtra utilities, CGPL, etc. The main argument advanced by the consumer organizations was that increase in the price of coal which makes the contract commercially onerous to perform cannot become the basis for frustration of contract.

The Supreme Court had called for the views of Government of India before deciding on the appeal. GoI was represented by Attorney General of India who made the following submission, extracted from the Supreme Court order dated 11 April 2017[6]:

> The electricity sector, having been privatized, has largely fulfilled the object sought to be achieved by the 2003 Act, which is that electricity generation, being delicensed, should result in production of far greater electricity than was earlier produced. He urged the court not to disturb the delicate balance sought to achieved by the Act i.e. that producers or generators of electricity, in order that they set up power plants, be entitled to a reasonable margin of profit and a reasonable return on their capital, so that they are induced to set up more and more power plants. This must be consistent with competitiveness among them, which then translates itself into reasonable tariffs that are payable by consumers of electricity. For

[6] SCC Civil Appeal Nos. 5399–5400 of 2016 dated 11th April 2017.

> this purpose, he relied strongly upon Section 3 of the Electricity Act, which states that the Central Government, shall from time to time, prepare a National Electricity Policy and a tariff policy in consultation with the State Governments, and the authority for development of the power system, based on optimal utilization of natural resources … He also referred us to the Cabinet Committee for Economic Affairs recognizing the overall shortfall in manufacture of domestic coal and the new coal distribution policy issued in July, 2013 pursuant to the Cabinet Committee which, according to him, are in the nature of binding directions making it clear that as generators of electricity, who depend upon indigenous coal, have been given less coal than was anticipated, should be allowed either to import the coal themselves, or purchase imported coal from Coal India Ltd., with the difference in price being passed through to them. He further referred to and relied upon the revised tariff policy of 28th January, 2016 for the same purpose. (ibid.)

From the Attorney General's submissions, it is evident that the government supported passing through of imported coal costs into the tariff since Coal India was not in a position to supply the entire contracted quantity to power developers and the shortfall had to be made good through imports. But the government counsel did not make clear the government's stand on whether the increase in coal prices on account of Indonesian notification should also be passed through in the tariff.

As for applying Force Majeure to the current context, the Supreme Court made the following observations:

> the doctrine is not to be lightly invoked; that mere incidence of expense or delay or onerousness is not sufficient; and that there has to be as it were a break in identity between the contract as provided for and contemplated and its performance in the new circumstances … It is clear from the above that the doctrine of frustration cannot apply to these cases as the fundamental basis of the PPAs remains unaltered. Nowhere do the PPAs state that coal is to be procured only from Indonesia at a particular price. In fact, it is clear on a reading of the PPA as a whole that the price payable for the supply of coal is entirely for the person who sets up the power plant to bear. The fact that the fuel supply agreement has to be appended to the PPA is only to indicate that the raw material for the working of the plant is there and is in order. It is clear that an unexpected rise in the price of coal will not absolve the generating companies from performing their part of the contract for the very good reason that when they submitted their bids, this was a risk they knowingly took. We are of the view that the mere fact that the bid may be non-escalable does not mean that the respondents are precluded from raising the plea of frustration, if otherwise it is available in law and can be pleaded by them. But the fact that a non-escalable tariff has been paid for, for example, in the Adani case, is a factor which may be taken into account only to show that the risk of supplying electricity at the tariff indicated was upon the generating company.

The Supreme Court also came unequivocally to the conclusion that Change in Law in the PPA referred only to change in Indian law. However, it did allow costs arising from change in Indian law to be passed on to the procurer of power through supplementary PPAs.

> … in so far as the procurement of Indian coal is concerned, to the extent that the supply from Coal India and other Indian sources is cut down, the PPA read with this document provides, in clause 13.2, that while determining the consequences of a change in law, parties shall have due regard to the principle that the purpose of compensating the party affected by such change in law is to restore, through monthly tariff payments, the affected party to the economic position as if such change in law has not occurred.

Besides, the apex court pointed out that Force Majeure Exclusions contained (i) any event or circumstance which is within the reasonable control of the parties …

Unavailability, late delivery or changes in cost of the plant, machinery, equipment, materials, spare parts, fuel or consumables for the project do not constitute 'Force Majeure';

"We are, therefore, of the view that neither was the fundamental basis of the contract dislodged nor was any frustrating event, except for a rise in the price of coal, excluded by clause 12.4, pointed out. Alternative modes of performance were available, albeit at a higher price. This does not lead to the contract, as a whole, being frustrated. Consequently, we are of the view that neither clause 12.3 nor 12.7, referable to Section 32 of the Contract Act, will apply so as to enable the grant of compensatory tariff to the respondents".

As for Change in Law argument, the apex court held that change in Indonesian law would not qualify as a Change in Law, but change in Indian law, to the extent it reduced the average contracted quantity of coal available to the power plants, certainly would qualify for compensatory tariffs to make up for the additional costs incurred in importing such quantities.

The Supreme Court also held that CERC could have invoked Section 79(1)(b) only in a situation where there are no GoI guidelines framed at all or where the guidelines do not deal with a given situation.

Section 79(1)(b) lists the following as one of the functions to be discharged by CERC "*to regulate the tariff of generating companies other than those owned or controlled by the Central government … if such generating companies enter into or otherwise have a composite scheme for generation and sale of electricity in more than one state*" (ibid.).

Since GoI guidelines had already been framed and implemented in the extant case, the regulator could not exercise its regulatory jurisdiction under Section 79(1)(b).

After the claims of the three generators were thrown out by the Supreme Court on 11.4.2017, Essar stopped supplying power to GUVNL from December 2017, claiming it had eroded its net worth and owed the banks Rs. 5062 crores as of 9.6.2017. CGPL too claimed its net worth had been wiped out, and it had accumulated a debt burden of Rs. 6306 crores as on 25.4.2017. APMuL, which had also stopped supplying power from its Mundra project from January 2018, wrote a letter dated 16.4.2017 to GUVNL offering 51% of its equity at Re.1, claiming it was on the brink of bankruptcy. It had also offered that if coal was supplied by GUVNL, it would supply power at capacity charges only (GUVNL Petition to GERC, 2019).

On 20.06.2017, a meeting was held at New Delhi by the Ministry of Power, Union of India of all stakeholders including representatives from the State Governments of Gujarat, Maharashtra, Punjab, Haryana, Rajasthan, representatives from various Banks and Financial Institutions (PFC and REC) and representatives from generating companies for resolution of the issues being faced by imported coal-based power plants. Pursuant to the said meeting, a Working Group was constituted of members of all the procurer states, and banks represented by Punjab National Bank and Canara

Bank with the SBI acting as a convener to find a solution through a consultative process.[7]

The Working Group Committee submitted its report on 10.01.2018 after undertaking due diligence on technical, legal and financial aspects of these three imported coal-based power projects in Gujarat. The Working Group Committee inter alia recommended two options (ibid.).

(i) state Govt. to takeover projects through majority stake and subsequently PPAs to be amended to address viability issues by allowing fuel cost pass-through or
(ii) amending PPAs without change in ownership—wherein procurers may agree for pass-through of fuel cost.

Meanwhile, SBI, the lenders to APMuL, also requested the formation of a High Powered Committee comprising experts from judiciary, banking and regulation to review the working group's report and recommend solutions.

In February 2018, the Gujarat government approached the centre once again on the basis of the Working Committee report and in April 2018 received the permission of the central government to take the lead on finding a solution, as it was the biggest procurer of power from the three power plants. Accordingly, the Gujarat government constituted the HPC in June 2018. It comprised Justice R. K. Agarwal, a former judge of the Supreme Court, Mr. S. S. Mundra, former Deputy Governor of RBI and Dr. Pramod Deo, former Chairman of CERC to go into the complaints and find a resolution.[8]

3.3 Essar

Essar Salaya, another 1200 MW coastal project in Gujarat that had come up through the competitive bidding route under Bid-3 around the same time as CGPL and APMuL was also entirely based on imported coal. It had signed a PPA with GUVNL to supply 1000 MW on 26.12.2007. It had commenced commercial operation from 2012, but had also been impacted by the Indonesian coal notification which took effect from 2011. Essar had stopped supplying power to GUVNL. In 2014, Essar had filed a petition similar to CGPL and APMuL, but with GERC (since it supplied to only Gujarat) seeking relief through supplementary tariffs (1389 of 2014), but which had not been decided by the regulator. On 9.6.2017, it had also informed GUVNL that its cash flows had dried up and its losses amounted to Rs. 5062 crores. As such, Essar claimed it was not in a position to supply power unless granted supplemental PPA to compensate for the increase in coal prices. Essar's demand for supplemental tariff was also referred to the same expert committee which was looking into the claims of CGPL and APMuL. Thus, the HPC formed on the recommendation of the

[7] CERC order dated 12 April 2019 in Case No: 374/MP/2018.

[8] https://www.livemint.com/Industry/WxS658YTKjcCMlDlA14U7I/Gujarat-govt-forms-committee-to-help-troubled-power-plants.html accessed on 16th May 2023.

Working Group addressed claims of all the three Gujarat-based projects which were affected by Indonesian notification.

The terms of reference of the HPC were

(i) The committee should examine and analyse all the relevant documents related to the projects.
(ii) Analyse and ascertain any hardship faced by these projects—if yes, mode and manner for mitigating the hardship faced by the projects on account of Indonesian Regulations and subsequent orders/judgements in the matter.
(iii) Call relevant parties for submission/details/clarification as required by the committee.
(iv) Contribution by each stakeholders, project developers and procurers concessions for mitigating hardship.
(v) Any other relevant issues which committee would like to discuss and study.
(vi) Suggest sustainable solutions for resolving the issue.
(vii) The committee may suggest any other measure for overall reduction in the cost of generation of power in the interest of the consumers.

The HPC met company officials, discom and state government officials, bank representatives and consumer representative groups, and sought their views and suggestions on how to proceed. Additionally, it took into account data analysis that suggested that despite the increase in cost of Indonesian coal, electricity generated at these power plants remained a cheaper option for the procurer states than other alternatives. Having taken all these views into account, the HPC's draft report has laid out its recommendations.

The committee recommended the following:

- The fuel cost risk should be borne by the procurers and eventually by the end consumers.
- This increased tariff would be limited to a cap of USD 110/tonne of coal and that any hike above that limit would have to be borne by the project developer.
- The lenders too would bear part of the increase in tariff to the extent of Rs. 9215 crores which would bring down the fixed tariff by 20 paise per kWh.
- The developers too would bear their share of the pain as the losses that they incurred on the power plants would not be fully compensated.
- The HPC suggested that its recommendations be implemented from 1 October 2018 and that no retrospective changes to tariff or 'compensatory tariff' can be entertained.
- Past losses to the tune of Rs. 21,249 crores (Adani—9473, CGPL—8176 and Essar—3600) will be to the account of the developers.

"Tariff will be adjusted considering actual fuel cost based on the superior or actual parameters or normative parameters … subject to a ceiling of HBA Index of 110 USD/mt (6322 kcal/kg, GCV) and the generator shall bear the differential cost and continue to supply power. Further, it is decided to review the ceiling price every five years as per the HPC's recommendations …" 100% of mining profits towards coal utilized at respective projects stipulated from their coal mines shall be shared. This will amount to 5 paise/kWh for Adani and 15 paise/kWh for CGPL, plant availability to be increased from 80 to 90% at no additional cost to procurers and availability below 90% to be penalized.

The committee claimed its recommendations were based on the principle of 'balancing of equities', whereby the pain is shared by all stakeholders. HPC claimed that the 'bona fide on the part of the developers to continue with the projects' was apparently demonstrated by the fact that they had already incurred higher losses than the damages that they would have had to pay if they chose to break the contracts. On the possibility of a resolution through the insolvency process at the NCLT, the report throughout appears to consider it a priority to prevent this eventuality—that is, to save the companies from being dragged down the NCLT route. The reasons adduced by HPC to come to its recommendations were as follows:

(i) The capacities from these projects have to be replaced from alternative sources, and therefore, prices will further go up in view of the clear co-relation between demand and supply.
(ii) The cost of replacement power at today's market price would be higher.
(iii) Setting up new projects will in any event be more expensive and will take 4–5 years to commence supply.
(iv) Increase in cost on account of procurement of power from inefficient and old plants which would also have reliability issue.
(v) Resorting to load shedding on account of difficulties associated with complete replacement of power from these projects was not desirable.
(vi) Any insolvency or liquidation of these projects would hardly address the issues of power supply.

In view of the above, the HPC recommended that PPAs be amended.

For GUVVNL, average variable cost would go up from Rs 1.74/kWh to 2.59/kWh. The capacities supplied by these plants to Western states are in the region of 16–18% of their total power supply and hence too substantial to be replaced by any other source.

To obviate any legal difficulty, Govt of Gujarat and GUVVNL approached the Supreme Court with Miscellaneous Application No. 2705–2706 of 2018 seeking clarification about impingement of Supreme Court order of 11.4.2017. Energy Watchdog, the consumer organization which had earlier appealed in Supreme Court, also approached the apex court with an appeal. Making CERC the respondent, Energy Watchdog prayed that it be heard before deciding on supplementary tariffs as recommended by HPC.

J. J. Rohinton Nariman, the same judge who had given the earlier judgement in the CGPL, Adani cases issued an order on 29.10.2018. This time, the apex court ordered that CERC expeditiously decides based on the HPC recommendations after hearing the objections of Energy Watchdog and other consumer representatives. Accordingly, CERC heard the case again and issued an order on 12.4.2019 allowing supplemental PPA to power plants using Indonesian coal, namely CGPL and APMuL. GUVNL approached GERC with a prayer that the Government of Gujarat notification dated 1.12.2018 accepting HPC recommendations should be made applicable also to Essar

Salaya in the light of the Supreme Court order dated 29.10.2018 pertaining to CGPL and APMuL with similar context. GERC granted approval of supplemental PPA between GUVNL and Essar after hearing Energy Watchdog and Prayas who were also impleaded in the petition as respondents. The PPA signed between the parties dated 01.03.2019 was ordered to be modified to accommodate review of coal price, ceiling price of coal, SHR, Auxiliary Consumption, incentive in excess of 90%, etc., for determination of energy charge, etc., through a supplemental PPA, much on the same lines as the CERC order on CGPL and APMuL.

Pursuant to the aforesaid policy decision and the Supreme Court's order dated 29.10.2018, APMuL and GUVNL mutually agreed to amend the PPA dated 2007. GUVNL filed Petition No. 374/MP/2018 before the commission seeking approval for the proposed amendments to Bid-01 PPA and Bid-02 PPA in terms of Article 18.1 of the respective PPAs. The said amendments were approved by the commission vide order dated 12.4.2019. Supplemental tariffs were allowed from 15.10.2018, whereupon, APMuL through an invoice claimed the differential tariff for the period from 15.10.2018 to March 2019 from GUVNL in terms of SPPA. Subsequently, APMuL kept claiming for differential tariff for each month as per provisions of SPPA.

GUVNL paid energy charges to APMuL for more than one year (in respect of invoices for months from October 2018 to November 2019) without raising any dispute. However, on 27.1.2020, GUVNL for the first time objected to pay APMuL certain components of tariff as per SPPA. From February 2020 onwards, GUVNL started making unilateral deductions against APMuL's monthly invoices of December 2019 and thereafter. On 15.4.2020, GUVNL intimated its following decisions to APMuL:

(i) GUVNL has decided to withdraw tolerance limit up to 10% allowed over HPB so as to align it with HPC recommendations and GR dated 1.12.2018.
(ii) GUVNL is restricting payment of energy charges to APMuL from October 2018 onwards considering coal price as lower of actual FOB price ...... APMuL to go for retendering for a prudent, competitive and cost-effective sourcing of coal in line with the coal price as per the market trend (**from Petition No. 614/MP/2020**).

In 2020, GUVNL filed a petition in CERC with the following prayers:

(a) Recall the order dated 12 April 2019 passed by the Hon'ble Commission in Petition No. 374/MP/2018 granting approval to the Supplementary Agreements {two} dated 05.12.2018.
(b) Declare that the two Supplemental Agreements dated 5.12.2018 entered into between GUVNL and Adani in regard to the PPAs dated 2.2.2007 and 6.2.2007 are void on account of the mistake as to the matter of fact and are, therefore, not enforceable.
(c) Declare that Adani has been in breach of the Supplemental Agreements dated 5.12.2018 the company having proceeded to claim and enforce the relief for the period from the date of the commercial operation of the generating units 5 and

6 of the Mundra Power Project till 14.10.2018 contrary to the provisions of the Supplemental Agreements.
(d) Direct Adani to refund the amount paid by GUVNL to Adani in terms of the Supplemental Agreements dated 5.12.2018 and also any further amount that may be paid to Adani by GUVNL till the adjudication of the present petition before this Hon'ble Commission with interest at the rate of SBI PLR plus 2% per annum.

Meanwhile, the notification issued by Government of Gujarat on 1.12.2018 accepting the recommendations of HPC with certain modifications was revoked on 12.6.2020. Yet another notification issued on the same date stated that the supplemental PPAs of CGPL and APMuL would be decided on a case to case basis.

Even as CERC was hearing GUVNL's petition #374/MP, GUVNL had approached the Supreme Court with a curative petition in 2020 demanding revisiting the earlier judgement of the apex court on reopening PPAs. GUVNL was seeking reopening of the apex court judgement on points of law as well as facts. The curative petition of GUVNL argued that the Supreme Court judgement given by J. J. Arun Misra did not take into consideration all the facts presented to it to show that the Adani Mundra project was not predicated upon coal supply from Morga block (see Chap. 5). A curative petition is rarely admitted by the Supreme Court, but in this case, the Chief Justice set up a full bench to hear and decide the petition anew, stating that there were substantial points of law and facts. The court had also served notice to the respondents accordingly.

However, in February 2022, the Government of Gujarat which got itself impleaded in the petition and sent the Attorney General to inform the Supreme Court that both GUVNL and Adani had reached a settlement and as such requested withdrawal of the case. The court allowed the withdrawal of the petition.

While the settlement deed is not in the public realm, according to court records Adani had agreed to forgo its claims on past dues and would settle only for fuel cost pass-through with effect from 15.10.2018. As per the Settlement Deed, the fixed charges as per the PPA dated 2.2.2007 read with SPPA dated 15.10.2018 shall be payable without any further change for the entire duration except for the period from 10.7.2019 till resumption of power supply. The Settlement Deed further provides that for energy charges, there shall be pass-through of cost of coal in a prudent and transparent manner and shall be linked to the escalation rates notified by the commission from time to time for which base rate shall be determined by the commission as on 15.10.2018 based on the normative parameters as per the provisions of the SPPA for the three parameters, namely (a) FOB of coal cost (USD/kWh), (b) Ocean Freight (USD/kWh) and (c) Port Handling Charges (Rs./kWh). Besides, according to the Settlement Deed, Adani agreed to relinquish its demand for termination of PPA and forgo all past dues. CERC will now determine the base fuel charges to which will be applied escalation rates as determined by the commission.

3.4 Analysis

The section above traced, in detail, the tortuous process of litigation with respect to three imported coal-based thermal power projects, all located in coastal Gujarat, two of them supplying to multiple states. Even after meandering for 15 years through a veritable labyrinth of litigation, res judicata eludes these projects; fresh appeals are being filed by new stakeholders even now, as this book goes to print.

The triggers for the protracted litigation are not far to seek. The financial stakes for either party are substantial, the consequences of contract non-performance, disastrous for consumers faced with prolonged power cuts; the hapless utility too is affected, forced to scramble for alternatives to the failed plants to maintain supplies including buying expensive power from real-time markets. Being an essential commodity, basic to human life as well as a critical driver of the economy, prolonged power cuts are not really an option.

But the deeper and more troubling questions that come to the fore in the plethora of disputes, petitions, appeals and counter-appeals, relate to the concept of the 'sanctity of contracts'. The objective of this analysis is not to pronounce on the merits or otherwise of the arguments made by the stakeholders and/or the validity or the quantum of compensation sought or sanctioned, although to the extent they impinge on the larger context of public interest, they find mention in the analysis.

The three Mundra projects alone have generated numerous disputes so far, both in regulatory commissions and in the Supreme Court. The costs of such litigation will have to be borne eventually, either by the rate-payer in the form of substantial tariff hikes or by the tax-payer through devolution from the state budgets. If the former, it will entail steep and avoidable increases in the tariff; if the latter, resources which could have been used in priority social sectors such as health or education would be diverted to fund protracted litigation with the major beneficiaries being the legal community. A developing country like India can ill-afford such a diversion of resources, especially when the disputing discoms themselves are already in a parlous state of financial health, having toted up accumulated losses of Rs. 100,932 crores as of February 2022.[9] The higher judiciary in India has a backlog of several million cases pending before it, yet a substantial chunk of its time has been pre-empted by this protracted high-profile, avoidable litigation. Similarly, regulatory agencies, tasked with fairly onerous regulatory responsibilities such as tariff determination or regulating quality of service, have had to divert enormous resources and time to adjudicatory processes as a result of the litigation.

At the beginning of this chapter, we asked what are the circumstances that lead to demands for reopening of contracts. This case study highlights the enormous complexities involved in writing time-consistent contracts for long-duration infrastructure projects. It is impossible to foresee all the eventualities that could frustrate the performance of a contract during its long lifetime.

---

[9] Reply to Parliament Question # 2240 Rajya Sabha given on 22.3.2022.

We also raised the question as to whether the assumption that governments/government entities always act in the consumer interest is valid. The foregoing analysis does not bear out this assumption, since the discoms do not have the muscle to hold developers down to their commitments. Despite corporatization, discoms are not commercially savvy yet. Already cash-strapped, the discoms are unable to pay the power suppliers on time which puts them on a weak wicket. And when Gujarat discom was ordered to pay supplementary tariffs to the project developers to make up for escalation in imported coal costs both by the regulators and APTEL, it did not even escalate its grievance to the apex court, although the increased tariffs would seriously impact its own finances and passing them on to consumers in tariff increases would be contentious. It was left to two consumer organizations to contest the CERC order in APTEL and again in the apex court. It was only subsequently, in 2021, that GUVNL filed a curative petition in the apex court, but withdrew it abruptly, claiming it had reached a settlement with the developers. The government too intervenes, setting up committees to advise on the quantum of compensation rather than allowing due process to take its own course and a market economy to find its own equilibrium.

As for the regulator championing the consumer interest, there is limited evidence that consumer interest features in regulatory adjudication. The analysis also glaringly highlights the absence of consumers as stakeholders in regulatory proceedings. It was only at the later stages of litigation when the regulators issued orders approving the demand for supplementary tariffs and the discoms failed to challenge it through appeals that two NGOs representing consumers—Prayas and Energy Watchdog—entered the fray, escalating the dispute to APTEL and again to the apex court. It is just a fortuitous coincidence that the states that would be most affected by the disputes happened to have NGOs with years of experience in championing consumer interests. Most states do not have active consumer organizations who can intervene in regulatory proceedings. Yet, there is little evidence of regulators grooming consumers to participate effectively in regulatory proceedings. The omission of a single major stakeholder—one who pays for the electricity—is egregious in a regulatory paradigm of governance.

While the project developers come armed with an impressive galaxy of legal luminaries to argue their case, consumer organizations were often left to depend on their own slender resources, pointing to serious power asymmetry among stakeholders. This is quite contrary to the working of regulatory commissions in the USA and elsewhere where consumer advocacy departments are part of the regulatory set-up and are financed and trained by the regulatory commissions themselves. The staff of consumer advocacy departments intervene in regulatory hearings on behalf of the consumers.

There also seems to be much divergence in the interpretation of the provisions of the statute as well as the terms of the PPA by the various adjudicating authorities. CERC and APTEL came to the rescue of the project developers, but on differing interpretations of the contract. The Supreme Court came to an entirely different interpretation and rejected the demand for supplementary tariff, deeming that sanctity of contracts was inviolable. Yet, the government decided to set up a committee to determine the terms of the supplementary tariffs on grounds of practicality. Such glaring divergences in the interpretation of contracts calls for introspection and harmonisation.

The unmistakable and unfortunate message to potential investors ensuing from the contradictory and often conflicting interpretation of the contracts by a spectrum of adjudicators—regulatory agencies, APTEL, high courts and the Supreme Court of India—is that commercial contracts in the Indian power sector are indeed amenable to diverse interpretations. The lessons emerging from the Mundra disputes will not be lost on future investors.

While information asymmetry is a perennial problem in large infrastructure projects, the discoms cannot absolve themselves of the responsibility to conduct due diligence of the bids they receive in order to ensure that tariffs quoted are within the realistic range. In their eagerness to secure the lowest tariffs for their consumers, discoms often hasten to award the contract without verifying the feasibility of such bids. Caveat Emptor should inform discom decisions which should, of course, be made in a transparent manner.

Besides, discoms should endeavour to hold the developer down to its commitment. After all, the developer took a commercial risk in anticipation of rewards, and when it goes awry, the developer must be made accountable to bear the costs of such risks.

The bidding guidelines allow a wide berth to the bidder to choose from among multiple options, perhaps on the premise that market knows best and therefore will come up with optimal bids factoring in all the risks. But as it turns out, such wide choices allow ample room for glossing over risks not readily apparent at the bidding stage, not to mention the scope it allows for manipulation in order to win the bid. The lesson that emerges from the Mundra disputes is that the government would do well to narrow down the choices available to the bidders, make them watertight, yet build-in sufficient safeguards to take care of contingencies that cannot be anticipated or envisaged ab initio. Both bid documents and PPAs must be detailed enough to avoid ambiguities and uncertainties which have become a playground for skillful lawyers.

There does not seem to be coordination between the stakeholders which could have obviated some of the obstacles that arose in the performance of the contract. For instance, coal allocation policies seem to have undergone changes post-FSA without regard to commitments already made to power plant developers. It is especially critical to insert some degree of finality into contracts especially because these are large volume projects that supply basic utilities such as electricity. 'Too big to fail' cannot become an excuse to allow violation of the sanctity of contracts.

Similarly, any provisions in the contract involving potential areas of conflict of interest will have to be minutely scrutinized and all questionable issues arising therefrom, settled ex-ante before factoring them into the contract. For instance, both the litigants in the present instance own coal mines in Indonesia through their subsidiaries and proposed to use coal from these mines in the power project. In a competitive bid, the bidder does not have to disclose the components of tariff bid by them including the assumed price of coal from own mines abroad. But as the dissenting note of a member of CERC pointed out, where the benefits of the tariff increase sought by the litigants accrue to their own overseas affiliates through transfer pricing, ex-ante clarity expressly included in the contract could have avoided vexatious and self-serving claims.

While ultra-mega power project policies seek to harness scale economies, the flip side is the gargantuan size of the projects which, when things go wrong, have cascading deleterious impact on a large set of stakeholders—consumers, discoms, ancillary industries, employees and indeed the nation's energy security. Perhaps the time has come for revisiting such large-scale projects and splitting them into manageable units as is happening in the nuclear power sector where small modular reactors are replacing large ones. In fact, as renewables play an increasingly salient role in the energy mix, backing down smaller conventional units to accommodate renewable power is less costly than backing down large thermal projects (see Chap. 5) as is happening now.

The question whether pragmatism should trump sanctity of contracts can never be conclusively argued. As long as legal provisions safeguard contracting parties' rights, a legal solution will be sought. While watertight legal provisions and detailed contracts that minimize ambiguities should guide the competitive bidding processes, efforts must always be to reach a settlement through negotiation in the event of disputes. This will avoid wastage of regulatory and judicial resources apart from astronomical legal fees all of which devolve on the consumers eventually. Even in this particular case, Adani, one of the litigants, eventually settled for a negotiated settlement with Government of Gujarat, but did so after the business environment was already vitiated by the prolonged and cascading litigation.

Most infrastructure industries are a complex web of legally enforceable contracts, requiring a degree of expertise which is why independent regulatory agencies comprising professionals with expertise were set up in the first place. In fact, independent regulatory agencies, which are not elected bodies, have still found their legitimacy in the larger Madisonian system of checks and balances, by virtue of their expertise. Therefore, ideally, disputes arising in the sector should find closure at the expert level, i.e. level of regulatory commissions rather than escalating them to civil courts. But India's electricity regulators have not always displayed expertise or maturity in dealing with complex problems arising in the sector. Article 226 of the Indian Constitution is wide enough to allow writ jurisdiction on any matter and litigants in the energy sector have not shied away from invoking this provision, whether warranted or not. This research concludes that it is up to the judiciary to admit or dismiss the petitions based on their salience in a larger jurisprudential context and the degree of expertise necessary for adjudication. Perhaps the judiciary can exercise this choice in a manner that deters vexatious litigation.

3.5 Chronology of Cases Relating to Sanctity of Contracts

3.5.1 Adani Power Mundra Limited (APMuL)

1. 21.2.2022—CERC (275/MP2021) in Adani vs GUVNL, taking on record, the settlement reached between the parties on 3.1.2022, and dismissing as infructuous, petition #275/MP for determination of tariff and accepting APMuL's relinquishment of past claims. Petition to be filed by APMuL/GUVNL for determination of pass-through coal costs.
2. 8.2.2022—Supreme Court allows withdrawal of curative petition filed by GUVNL against Adani on the Review petition # 2012 of 2019.
3. 28.6.2021—CERC order (614/MP/2020) Adani vs GUVNL on making unilateral deductions in tariff payable. Ordered to pay full tariff without deductions.
4. 28.5.2020—CERC Record of Proceedings in Petition No. 250 and 275 of 2009 in CERC.
5. 2.7.2019—Supreme Court order in Adani vs GERC (Arun Misra order on linking Mundra to coal allocation).
6. 12.4.2019—CERC (374/MP) order on GUVNL vs Adani, Prayas, Energy Watchdog on implementing Supreme Court order for supplemental tariffs based on HPC.
7. 12.2018—APTEL (241 of 2018) Interim order on sending power to third parties before SCOD.
8. 29.10.2018—Supreme Court in Energy Watchdog vs CERC (Nariman allowing CERC to decide on HPC recommendation).
9. 11.4.2017—Supreme Court in Energy Watchdog, Prayas et al. vs CERC (Nariman dismissing Change in Law and FM).
10. 7.4.2016—APTEL full bench (100 of 2013 and 98 of 2014) order on compensatory tariffs under Force Majeure.
11. 6.1.2016—CERC (155/MP2012) in Adani vs multiple discoms, Prayas allowing compensatory tariffs under special powers subject to outcome of appeal (5399–5400 of 2016) in the Supreme Court.
12. 9.10.2015—CERC (24/MP/2014) Adani vs GUVNL demanding compensation for exchange rate variation and tax exemption.
13. 4.10.2012—CERC (374/MP) GUVNL vs Adani/Prayas/EW—on scheduled commercial date of delivery (not really relevant to the chapter).
14. 7.9.2011—APTEL (184 of 2010) Adani vs GERC et al.—termination of PPA illegal; directing Adani to continue to supply power and its project not linked to FSA with GMDC.

3.5.2 Coastal Gujarat Power Limited (CGPL)

29.10.2018—Supreme Court order in Energy Watchdog vs CERC (Nariman allowing CERC to decide on HPC recommendation).
June 2018—Government of Gujarat appoints High Powered Committee.
11.4.2017—Supreme Court order in Energy Watchdog, Prayas et al. vs CERC (dismisses Change in Law and Force Majeure).
7.4.2016—APTEL order on CGPL appeal against rejection of 'Force Majeure' by CERC.
21.2.2014—CERC order on compensatory tariffs.
16.8.2013—expert committee submits report to CERC.
1.4.2013—CERC order on CGPL petitions. Expert committee set up to determine compensatory tariff to CGPL and AMPuL on the recommendation of CERC.
July 2012—CGPL files petition in CERC against GUVNL, MSEDCL, Ajmer Vidyut Vitaran Nigam Ltd., Jaipur Vidyut Vitaran Nigam Ltd., Jodhpur Vidyut Vitaran Nigam Ltd., PSPCL, HPGCL and Union of India.

3.5.3 Essar

2014—Files for supplementary tariffs in GERC.
9.6.2017—Stops supplying power to GUVNL.
Clubbed with AMPuL and CGPL.

References

Demsetz, Harold. 1968. The Cost of Transacting. *The Quarterly Journal of Economics* 82 (1)(Feb., 1968): 33–53 (21 pages). Oxford University Press.

Friedman, Milton. 1968. The Role of Monetary Policy. *American Economic Review* LVIII (1).

Guasch Luis J, Jean-Jacques Laffont, and Stephane Straub. 2022. Renegotiation of Conncession Contracts in Latin America, SC Center for Law, Economics & Organization, Research Paper No: C02-22.

Kydland, Finn E., and Edward C. Prescott. 1977. Rules Rather than Discretion: The Inconsistency of Optimal Plans. *Journal of Political Economy* 85 (3): 473–491.

Mehta, Abhay. 2000. *Power Play—A Study of the Enron Project*. Orient Blackswan.

Prayas Energy Group. 2011. *Prayas Discussion Paper on Capacity Addition Through Competitive Bidding in India Private Sector Participation in Generation: From MoUs to Competitive Bidding*, March 2011.

Chapter 4
Credible Commitments

This chapter deals with a cost-plus MoU-based project which meandered through a labyrinth of obstacles reflecting the absence of credible commitments on the part of the government, the government-owned discom as well as the project developer. Credible commitments are imperative for a successful transition to a market economy.

4.1 Hinduja National Power Corporation Limited

In 1992, India's power generation sector was opened up to private and international investors. The Independent Power Producers (IPP) policy offered very generous incentives and a two-part tariff policy to potential investors, exempting them from international competitive bidding and allowing cost-plus tariffs. The first eight IPPs to invest in India were termed 'fast-track' projects whose payment security was guaranteed by the state government and counter-guaranteed by the Government of India. The infamous and ill-fated Enron power project was the first to make an entry into India through this route.

One such 'fast-track' project, a coal-based 1040 MW thermal power plant to be set up in Visakhapatnam in the erstwhile state of Andhra Pradesh, was proposed by the London-based Hinduja group. The Government of Andhra Pradesh, through its integrated electric utility APSEB, signed an MoU with the Hinduja group in 1992. In 1996, Hinduja National Power Company Limited (HNPCL)'s proposal received techno-economic clearance by Central Electricity Authority which approved a capital cost of Rs. 4628.11 crores. In the same year, APSEB signed a PPA with HNPCL for the 1040 MW plant. In 1998, both parties mutually agreed to amend certain provisions of the PPA. The amended and restated PPA was signed on 15.4.1998.

S. Mahalingam and K. Mahalingam, *Independent Energy Regulation in a Developing Economy*, https://doi.org/10.1007/978-981-99-4398-2_4

However, differences arose between Government of India and HNPCL on the terms of the counter-guarantees offered by the former. Thereafter, the HNPCL project proposal went into cold storage for nearly ten years.[1] Subsequently, in the year 2007, HNPCL approached the Government of Andhra Pradesh to revive the power project, proposing to restructure it as a merchant plant, offering 25% of the power generated to the State on the basis of first refusal and balance 75% power to third parties. However, after negotiations between the parties, the state government had offered to purchase 100% of the power generated from the plant of HNPCL. HNPCL confirmed its agreement in writing.

Unrelated to this, in 2011, the Central Power Distribution Company (APCPDCL), a holding company of four Distribution Companies of Andhra Pradesh (APDISCOMS) had initiated a competitive bid for long-term power procurement of up to 2000 MW. The bids were called under Case1 of GoI guidelines wherein the bidder chooses the location, fuel, etc., and is also responsible for fuel procurement. The power from the winning project was expected to meet the base load requirements of APDISCOMS from the years 2014 to 2015 onwards.

HNPCL participated and had successfully emerged as the second lowest bidder (L2 bidder) during this bidding round. However, the Bid Evaluation Committee in its meeting dated 28.9.2012 had noted that the state government had already informed that the entire capacity of HNPCL was encumbered to the State of AP/APDISCOMS and was not available for consideration under the bid. Accordingly, the Bid Evaluation Committee had discarded HNPCL from the bidding process.

Meanwhile, in 2012, there was a correspondence between HNPCL and the state government regarding the steps to be taken for development of the HNPCL project. In August 2012, in a letter addressed to the then Chief Minister of Andhra Pradesh, HNPCL had requested state support for scheduled commissioning of the project. In December, 2012, the state government addressed a letter to HNPCL accepting its proposal and agreeing to purchase 100% power from the project of HNPCL as per 'the Amended and Restated PPA of 1998'. In a communication dated 14.1.2013, HNPCL agreed to supply 100% power to the State Distribution Companies at the tariff to be determined by the State Commission.

In May 2013, HNPCL wrote to the discoms, claiming that its capital costs of the original MoU project had gone up from Rs. 5545 crores to Rs. 6098 crores. The procurer discoms did not dispute this claim, but reserved their right to contest it before the APERC as and when the project came up for regulatory approval and tariff determination. On the very next day, HNPCL and the discoms signed a Memorandum of Agreement to proceed with the earlier PPA of 1998 on the same terms as in the 'amended and restated' version. Based on this MoA, HNPCL signed a Fuel Supply Agreement in August 2013 for supply of coal from Mahanadi Coal Fields.

In March 2014, HNPCL approached APERC for tariff determination. Even as APERC was considering the tariff application of HNPCL, on 2.6.2014, the Andhra Pradesh State Reorganization Act 2014 came into effect bifurcating Andhra Pradesh

[1] https://www.business-standard.com/article/economy-policy/hinduja-power-project-in-ap-looks-stillborn-101011701079_1.html, accessed on 23 March 2023.

into two states—the State of Andhra Pradesh and the State of Telangana. This necessitated re-assigning of the power project to the two AP discoms (as distinct from Telangana discoms)—AP Southern Power Distribution Company Limited and AP Eastern Power Distribution Company (discoms).

On 11.1.2016, 18 years after the PPA was first signed between the erstwhile APSEB and HNPCL, the first unit of the Power project (520 MW) was declared commercially ready to supply.

Meanwhile, on 28.7.2015 (O.P# 21 of 2015), HNPCL had filed an addendum with APERC, claiming increase in the project costs to Rs. 8087 crores. In its addendum, HNPCL claimed that it was going by the benchmarks for capital costs computed by CERC in its order dated 4.6.2012. This capital cost was disputed by the AP discoms. On 28.4.2016, the successor discoms of Andhra Pradesh—APSPDCL and APEPDCL—took over from the erstwhile APEDCL. On 1.6.2016, the newly bifurcated Andhra Pradesh state government accorded approval for purchase of 100% power from HNPCL as per the amended and revised terms per PPA signed in 1998. The Telangana discoms, part of the original arrangement prior to division of the state, had expressed their desire to offtake 54% of power from HNPCL for their new state, but Andhra Pradesh decided to absorb all the power from HNPCL for its own consumers.

The newly formed AP discoms filed a petition with APERC for approval of HNPCL PPA and determining tariffs (O.P.# 19 of 2016). On 6.8.2016, APERC determined the provisional tariff of HNPCL at Rs. 3.82 per unit, payable by the discoms for the power supplied by HNPCL. The determination of final tariff was pending. Meanwhile, HNPCL had started supplying power to the AP discoms at the interim tariff of Rs. 3.82/kWh.

In 2018, perhaps aggrieved by what they considered steep tariff determined by the regulator, the discoms decided not to draw any more power from HNPCL. Accordingly, they had not included power from HNPCL in their ARR filed for FY 2018. On 31.1.2018, the discoms even withdrew their application for final approval of the PPAs by APERC as well as their tariff proposal. APERC allowed the discoms to withdraw their application. HNPCL's own application for approval of increased capital costs had thus become infructuous and was dismissed by APERC.

In March 2018, an aggrieved HNPCL approached APTEL with an Execution Petition, demanding that the discoms be ordered to execute the 15.4.1998 PPA and continue to schedule power from its plant.

HNPCL prayed that APTEL

"direct ...the two Distribution Licensees in the State of Andhra Pradesh to continue to implement the Power Purchase and Continuation Agreement/ Procurement Process Agreement which was followed prior to 31.1.2018 in the same manner as before and maintain the status quo as was prevalent on 31.1.2018" HNPCL also prayed that APERC be directed to appraise its revised project costs and determine two-part tariff in a time-bound manner".

The regulations of the State Commission consider the scheduling of the thermal power plants on merit order dispatch which is based on variable (fuel) charges.

Accordingly, the bifurcation of provisional tariff of Rs. 3.82/kWh had become necessary and urgent. Therefore, APTEL directed APERC to "get into details of the tariff of Rs. 3.82/kWh for its bifurcation into fixed charges and variable charges" within 15 days. I also added that the fixed charges/variable charges so determined would be provisional and would be subject to the regulator's final judgement/order.

In conclusion, APTEL ordered, "In view of our discussions as above and considering all the relevant aspects of the instant Execution Petition, we are of the considered opinion that the matter needs to be remanded to the State Commission… for immediately determining the fixed charges and variable charges of the Project based on the provisional tariff, so that the execution of the order dated 16.3.2018 of this Tribunal is carried out expeditiously".

So, HNPCL submitted that its total interim tariff of Rs. 3.82/kWh should be split into fixed and variable costs and claimed a variable cost of Rs. 2.68/kWh which should be considered for scheduling its power under Merit Order Dispatch. At Rs. 2.68/kWh variable cost, in 2018, HNPCL's variable cost was within the merit order list and as such, power from HNPCL was due for dispatch.

APTEL ordered that the revised and restated PPA signed on 15.1.1998 be honoured and since HNPCL's variable cost was low enough to be included in the merit order, the discoms should continue to buy the power. APTEL also ordered that the status quo prior to withdrawal of application to APERC for approval of PPA be maintained. APTEL directed APERC, the regulator to expeditiously determine the final tariff of HNPCL.

The discoms then moved a writ petition in the High Court challenging APTEL's order, but it was not entertained by the Andhra Pradesh High Court. Thereafter, the AP discoms approached the Supreme Court with an appeal to set aside APTEL's execution order.

In the Supreme Court, the AP discoms argued that their PPA with HNPCL had not been approved by APERC and as such was not valid and that APTEL had given an order of specific performance of a contract that was void ab initio. The discoms further argued that MoA dated 17.5.2013 and the Continuation Agreement dated 28.4.2016 were themselves contrary to the National Tariff Policy and the tariff regulations of APERC. As such, the direction by the APTEL, to continue to get the electricity supply from HNPCL, being contrary to the statutory provision, would not be tenable in law.

The discoms also disputed HNPCL's claim that it made the investment and built the plant only because of the assurance given by the state government that it would buy all the power from the plant, maintaining that HNPCL had completed 93% of its investments in the project even prior to the state government making such commitment. HNPCL's investments had gone up from initially estimated Rs. 4628.11 crores to Rs. 8087 crores, pushing up the cost of power purchase steeply. Buying such costly power would not be in the public interest since consumers would have to bear the cost through their bills. In any case, the discoms argued, being independent corporations, they were not bound by the commitment made by the State. Besides, discoms could not purchase power without the prior approval of the State Commission.

Counsel for HNPCL also argued that the discoms should not be allowed to withdraw their application for regulatory approval for the project PPA and tariff because it would frustrate the contract. Also, that HNPCL had participated in the competitive bid only because it wanted to run the plant as a merchant facility, but it was only because of persuasion and subsequent commitment by the state government to procure 100% of the power to be generated from the plant. HNPCL also pointed out that consequent to the reorganization of the erstwhile Andhra Pradesh state, Telangana offered to purchase 54% of its power, but was pre-empted by the successor Andhra State which opted to take all the 100% of power generated by HNPCL. The counsels for HNPCL alleged malafide on the part of the discoms which were now buying power from other generators at variable tariffs much higher than that offered by HNPCL.

The Supreme Court in its order dated 2.2.2022 rejected the discoms' appeal to stay APTEL's order on execution of the contract. The court held as untenable, the discom's argument that they were not drawing power from HNPCL on account of the high capital cost of the project because it would translate into high tariffs affecting the public interest. The apex court pointed out that while determining the tariff, the regulator would be guided by various factors required to be taken into consideration under the extant law and regulations.

The apex court held that withdrawal of an application would not be permissible when such a withdrawal amounts to frustration of a contract. It also held that the argument, that on account of increase of the capital cost of the project, the discoms would be required to purchase power at a much higher rate was not tenable because, while determining the tariff, the regulator would be guided by various factors as are required to be taken into consideration under the provisions of the law. In any event, the discoms have themselves reserved their right to contest the correctness of the cost on every component at an appropriate stage before the State Commission. That the State Commission, while approving the cost of the project and determining the tariff would be required to look into various factors and is not bound to accept the same.

"It is thus trite that, while considering grant of approval to the PPA, the State Commission will have to keep in mind the public interest. It will have to consider, as to whether the PPA, which is subject to approval, subserves the public interest. It will also be required to take into consideration, as to whether the terms agreed are fair and just while granting approval. While exercising power under Section 86(1)(b) of the Act of 2003, the Commission will have to regulate the price at which the electricity would be procured from the generating companies. Undoubtedly, while doing so, the Commission will be guided by the factors mentioned in Section 61 of the Act of 2003 and the Regulations concerning the same".

Castigating the regulator, the apex court said that the State Commission was wholly unjustified in dismissing HNPCL's application for approval of capital cost. "Even a generator can file for tariff. It could thus be seen that after investment of huge resources including the land belonging to the State, the project is complete and has become operational. The question, at this juncture, would be, whether to discard such a project is in the public interest or against it".

The Supreme Court observed that the discoms had been purchasing power at the rate of Rs. 3.82/kWh from HNPCL till 14 July 2020, but stopped thereafter although they were purchasing power from other generators at even higher cost.

"We ask a question to ourselves, as to whether public interest, which is so vociferously pressed into service in the present matter by the appellants—DISCOMS, lies in purchasing the power at the rate of Rs. 3.82 per unit from HNPCL or by purchasing it at the rate of Rs. 4.33 per unit from KSK Mahanadi. We strongly deprecate such a conduct of the appellants—DISCOMS, which are instrumentalities of the State. The appellants—DISCOMS, rather than acting in public interest, have acted contrary to public interest. For defying the orders passed by this court, we could very well have initiated the action against the officials of the appellants—DISCOMS for having committed contempt of this court, but we refrain ourselves from doing so".

It ordered the discoms to 'forthwith' start purchasing the power from HNPCL at the rate of Rs. 3.82 per unit as per the orders passed by the APTEL dated 16 March 2018 and by this court dated 21.8.2020 and directed APERC to expeditiously dispose of the petition seeking final determination of HNPCL tariff, in any case within the next six months.

This order of the Supreme Court was passed on 2.2.2022. APERC has been directed to determine the final tariff of HNPCL within six months of the order.

4.2 Analysis

HNPCL had a valid PPA signed on 15.4.1998 with AP discoms for offtake of 100% of power generated from its Vizag plant. Yet, the discoms repudiated the PPA through the back channel, by withdrawing their petition for regulatory approval and tariff determination of HNPCL plant; the discoms also stopped drawing power from the project developed entirely for their benefit and with their approval expressed through the PPA. This research agrees with HNPCL's claim of restitution of the PPA.

The dispute between HNPCL and the discoms is on the cost of power supplied by the former. Not all project delay and the consequent cost escalation claimed by HNPCL can be attributed to the bifurcation of the state. To the extent the delays were on account of the developer, cost escalation should be disallowed. The discoms had a forum—APERC—where they could have contested the escalation in project costs and the consequent impact on tariffs during the hearings for approval of HNPCL costs. Instead, the discoms resorted to other methods which were rightly termed improper by the apex court.

Both APTEL and the Supreme Court were right in upholding the validity of the PPA and directing the discoms to continue drawing power from HNPCL. Now the ball is in the regulator's court to fix tariffs based on the escalated project costs claimed by HNPCL.[2]

[2] In August 2022, after the manuscript was submitted, APERC determined the tariff of HNPCL. The regulator pared down capital costs from Rs. 7758 crores claimed by HNPCL to Rs. 5810 crores

The project, first finalized and sealed through a valid PPA in 1998, was completed only in 2016, entailing a delay of 18 years in its execution. This delay is primarily the cause of escalation in project costs claimed by HNPCL which, in turn, has given rise to the dispute between HNPCL and the discoms. It is curious that in July 2015, HNPCL filed an application with APERC for escalation of costs from Rs. 6098 crores to Rs. 8087 crores on the ground that it was aligning its costs with the benchmark capital costs for thermal power stations determined by CERC in its order dated 4.6.2012 instead of presenting details of actual cost escalation for the project. Without access to the details in the addendum filed by HNPCL, it is difficult to determine whether there was escalation in actual costs of setting up the plant or HNPCL was taking advantage of CERC order to claim higher capital costs.

In any case, it becomes incumbent upon the regulator to determine the reasons for cost escalation. If it turns out that cost overruns were on account of delays, the regulator must identify the parties responsible for delays and determine the tariff accordingly. Fixed tariffs will be payable through the 25 year life of the PPA to be passed on to the consumers. Public interest dictates that prudence should guide regulatory appraisal of capital costs and tariff determination. The CERC order clearly states that the regulator must apply the canon of prudence and necessity when examining capital costs and is not bound by the benchmarks announced by it. Thus, if the delay in construction of the project is attributable to the developer, the developer must bear the risk of cost overruns rather than foisting it on the consumer through tariffs.

It is also surprising that the erstwhile APSEB and the Government of Andhra Pradesh decided to drop HPNCL from the purview of a competitive bid conducted in 2011 even after the latter emerged as L-2. It is an acknowledged fact that competitively bid projects are superior to cost-plus projects because tariffs quoted in the bid are locked for the duration of the PPA (with admissible escalations, if any), whereas cost-plus projects always pose the risk of cost escalation. In projects of the magnitude such as HNPCL which not only entails substantial capital, but also many approvals, time overruns are a distinct possibility and with delays, cost overruns become ineluctable. Yet the Andhra Pradesh government and the discoms chose a cost-plus project over a competitively bid project, flouting norms of prudence and acting contrary to the public interest. If APSEB and AP government had hoped to tie down HNPCL to the 'amended and restated PPA' of 1998, they were certainly not realistic. The HNPCL case is a classic example of the absence of credible commitment on the part of state-owned discom and the AP government exacerbated by the developer's apparent lack of seriousness as evident from the inordinate delay in commissioning the project.

---

and fixed the variable unit tariff at Rs. 2.44/kWh as against Rs. 2.68/kWh allowed in the interim variable tariff applicable up to 2022.

4.3 Chronology in Hinduja National Power Corporation Limited

1. 2.2.2022—Supreme Court order on HNPCL
2. 2020—AP discoms appeal APTEL order in Supreme Court
3. March 2018—HNPCL approaches APTEL with petition to reinstate power drawal by AP discoms pleads for tariff split into fixed and variable. APTEL orders AP discoms to draw power and directs APERC to fix final tariff
4. 31.1.2018—Discoms withdraw tariff approval petition from APERC and stop drawing power from HNPCL
5. 6.8.2016—APERC assigns provisional tariff of HNPCL at Rs. 3.82/kWh
6. 28.4.2016—AP discoms bifurcated into AP and Telangana discoms. HNPCL PPA fully assigned to AP
7. 11.1.2016—First unit of HNPCL commissioned
8. 28.7.2015—HNPCL files addendum for additional capital costs with APERC
9. 2.6.2014—The Andhra Pradesh Reorganization Act 2014 comes into effect. State split into Andhra Pradesh and Telangana
10. March 2014—HNPCL files for tariff determination by APERC
11. May 2013—HNPCL claims escalated project costs
12. 14.1.2013—HNPCL confirms supply of power to APTransco at tariff determined by regulator
13. 28.9.2012—Evaluation Committee rejects bid by HNPCL
14. 2011—APCPDCL (Andhra Pradesh Central Power Distribution Company Limited, holding company of Andhra discoms) invites bid for 2000 MW power plant
15. 2007—HNPCL request for converting HNPCL into merchant plant. APTransco offers to buy all 100% of power generated by HNPCL
16. 14.4.1998—Amended PPA signed between APTransco and HNPCL
17. 1996—Central Electricity Authority's clearance for HNPCL based on MoU
18. 1992—Two-part tariff and GoI counter-guarantee announced for fast-track IPP projects.

Chapter 5
Interpreting/Implying Terms into Contracts

It is settled jurisprudence that a contract must be read as a whole and the intention of the parties must be gathered from the language used in the contract by adopting harmonious construction of all the clauses contained therein. Yet, on account of the complexities involved in long-term power purchase contracting and the plethora of rules, regulations, laws, amendments and notifications all of which have a bearing on the contract, regulators, APTEL and the Supreme Court have been called upon to imply terms into Power Purchase Agreements. Three case studies in this chapter examine how the regulator, APTEL and the Supreme Court have interpreted the terms of Power Purchase Agreements between private developers and the state-owned discoms.

Pacta Sunt Observanda (contracts are meant to be honoured) is an ancient dictum that maintains that people will make contracts only if they can rely upon their being enforced. Sanctity of contracts is a much-discussed principle in the context of a globalized economy. Transnational capital in search of investment opportunities in jurisdictions away from their home turf have sought watertight legal framework in host countries to protect their investments. WTO rules also champion a sound legal framework of which sanctity of contracts forms a core tenet.

Ever since India liberalized its electric supply industry in 1992 and invited private and foreign capital to invest in the sector, sanctity of contracts has assumed salience. Regulators, APTEL and the judiciary are routinely called upon to interpret terms of commercial contracts as more and more private investors are entering an industry hitherto dominated by state-owned enterprises. Contract interpretation invariably involves substantial sums of money for either contracting party and therefore, hotly contested.

In Chap. 3 we discussed how a chimera of sanctity of contracts was maintained even as innovative ways (supplementary tariffs) were found to tweak their sanctity. In this chapter, we examine whether contracts are inviolable even if they are too long-term to be time-consistent. Must they always be performed by both contracting parties (or pay liquidated damages for repudiation if the contract allows it) or are there circumstances where their sanctity gets diluted? If the latter, what are those

S. Mahalingam and K. Mahalingam, *Independent Energy Regulation in a Developing Economy*, https://doi.org/10.1007/978-981-99-4398-2_5

circumstances? What are the principles that should apply in deciding when a contract is sacrosanct and when it can be re-interpreted in the interest of pragmatism or practicality? Can the regulator or courts read into the contract something that is not explicitly stated? Should a contract be interpreted strictly in terms of the explicit meaning of the words, or should it be read to give it a normal business sense?

In this chapter, we seek answers to these and other questions. We examine how the regulator, the Appellate Tribunal for Electricity and the Supreme Court have interpreted three power purchase contracts and the principles they applied in such interpretation.

5.1 Nabha Power Limited Versus Punjab State Power Corporation Limited

5.1.1 Reddendo Singula Singulis

In 2009, Punjab State Electricity Board, the predecessor entity of Punjab State Power Corporation Limited (PSPCL) called for competitive bids for supply of 1040 MW power to the state of Punjab. The bids were called under 2005 Government of India Guidelines for Competitive Bidding Case 2 (wherein the project is both location and fuel-specific and the procurer of power assumes responsibility for ensuring fuel linkage) and Scenario 4 (wherein fuel costs based on actuals are pass throughs as energy charge).

The procurer had identified Rajpura as the location, allotted land, secured coal linkage from South Eastern Coalfields Limited (SECL), Chhattisgarh, and water supply for the plant. The bidder's responsibility included, among others, procurement and testing of coal. Prior to award of the contract, PSPCL had formed Nabha Power Limited (NPL) through a special purpose vehicle to be transferred to the successful bidder. L&T emerged the winner of the bid and acquired 100% shareholding in NPL and signed a 25-year Power Purchase Agreement with PSPCL.

Pursuant to Ministry of Environment and Forest Notification of 1997 as well as PSPCLs' prebid clarifications, NPL was bound to use washed coal.[1] The RFP issued by PSEB had included the following:

SECL would supply Grade 'F' coal from Korba/Raigarh field, with GCV of 3900–4260 kcal/kg, Ash Content of 35–40%, total inherent moisture of 5–6%, Volatile matter of 24–32%, fixed carbon of 32–37% and Sulphur content of 0.05%.

On a specific query as to whether the coal to be supplied would be washed coal or unwashed coal, it was clarified that washing of coal was to be arranged by the successful bidder (SCC Civil Appeal 179 of 2017).

In response to the queries raised by the bidders, clarifications on the model PPA were also issued on 17.09.2009. On the question of the costs associated with fuel

[1] Cited in Supreme Court order dated 5 October 2017 in civil appeal no. 179 of 2017.

supply, transportation and unloading being pass through, it was clarified that "tariff payment will be in accordance with Schedule VII of the PPA". Schedule VII lists the specific instances where extra-PPA costs would be allowed such as 'Force Majeure' or 'Change in Law', etc.

NPL commissioned the plant in 2014 and raised an invoice for payment of tariff for power supplied to PSPCL. However, even from the first invoice onwards, PSPCL made the following deductions from the amount payable to NPL.

(i) *Coal washery charges and cost of coal towards loss of quantity on account of washing (yield loss)*
(ii) *Gross Calorific Value (GCV) of coal to be taken for computation of energy charges should be on 'as fired' basis*[2]
(iii) *Road transportation cost—at the plant-end and at the mine-end*
(iv) *Liaising charges, transit and handling losses and third party handling charges*
(v) *Capacity charges for the period from 20.02.2014 to 03.03.2014 when the availability was declared on non-linkage (alternate) coal.*[3]

5.2 Punjab State Electricity Regulatory Commission (PSERC)

NPL approached PSERC with a dispute against PSPCL for prayer that all the above costs be reimbursed. However, PSERC dismissed the petition on 1.2.2016. The regulator gave the following reasons for the dismissal:

(i) *When coal supply committed by PSPCL was delayed, NPL used alternate sources to make the plant available at the committed capacity as per PPA, and this was inadmissible*
(ii) *Coal washing costs should have been factored in by NPL even at the time of bidding since the MoEF notification on using washed coal came prior to the cut-off date for bidding. It cannot be construed as 'Change in Law' contemplated in the PPA because the law was extant prior to the signing of PPA*
(iii) *Since coal washing costs were not admissible, loss of coal on account of washing is also inadmissible.*

---

[2] There is a significant increase in the energy charge with every 100 kcal/kg drop in GCV. With net heat rate of 2268 kcal/kWh and assumed cost of coal as ₹4000 per tonne, the energy charges per kWh work out to ₹2.16 with GCV 4200 kcal/kg and ₹2.39 with GCV 3800 kcal/kg. NPL claimed it was entitled to recover energy charges from PSPCL computed on the basis of 'as fired' GCV of coal in accordance with clause 1.2.3 of Schedule 7 of the PPA read with the CERC Tariff Regulations 2009. There is a substantial slippage in the GCV of coal due to reasons beyond the control of NP (PSERC order dated).

[3] Although coal linkage was PSPCL's responsibility, it could not arrange supply of coal in the first year of operation during which NPL made its own arrangements to procure alternate coal. The PPA had required NPL to be available at 80% capacity at all times to be eligible for fixed cost payments.

The regulator justified its decision on the ground that NPL came through the bidding process under Section 63 of Electricity Act 2003. As such, yardsticks available to cost-plus generating stations that come under Section 62 of EA 2003 could not be applied to NPL. PSERC further added that PSPCL bore no responsibility to get the coal washed at its cost. Moreover, PSPCL did not specifically agree to bear the cost of washing.

As for the GCV of coal sought by NPL 'as fired' and not 'as delivered', PSERC maintained that the rational interpretation of the relevant provisions in the PPA would be that PSPCL is liable to make payment of the coal cost component in the monthly energy charges on the basis of the bills for the cost of coal raised by SECL to NPL with GCV mentioned therein.

In respect of liaising, transport and unloading charges to be paid by PSPCL, the commission held that the relevant clause in the PPA did not cover the payment of such expenses separately, in addition to the cost of coal paid by NPL to SECL, transportation charges paid by NPL to railways and unloading charges provided for thereunder. The commission found there was no specific provision in the PPA for computing monthly energy charges considering any of the aforesaid charges nor did PSPCL, at any stage, agree to pay such charges to NPL separately. As per the commission's interpretation of the provisions in the PPA, such charges are not payable additionally by PSPCL.

5.3 APTEL

NPL then appealed to APTEL, reiterating the same claims. APTEL dealt with the following questions in detail

(a) *Whether the State Commission failed to act in accordance with its mandate as set out in the statutory framework comprising of Sections 61, 63 and 86(1) (b), (f) and (k) of the Electricity Act read with Clauses 4.7 and 5.17 of the Competitive Bidding Guidelines*
(b) *Whether the State Commission fell into error in holding tariff bid out under Section 63 of the Electricity Act as cast in stone which is beyond its jurisdiction?*
(c) *Whether the State Commission failed to discharge its statutory role in a competitively bid out PPA to adapt/adjust/provide for developments during the life of the PPA?*
(d) *Whether the PSERC order constitutes unlawful disallowances of legitimate claims and suffers from non-application of mind and lack of reasons insofar as several material submissions and legal principles argued by the NPL have either not been considered in entirety or have been dismissed without giving any reasons?*
(e) *Whether the State Commission erred in interpreting Clause 1.2.3 of Schedule VII of the PPA, since the said clause defines the cost of coal as the weighted average actual cost to the seller of purchasing, transporting and unloading the*

coal most recently supplied to and at the Project, which would mean that all actual expenses incurred in the process of bringing such coal to the project site would be included?

(f) *Whether the State Commission erred in disallowing payment of washing charges of coal to NPL, failing to appreciate that the cost for washing coal is a part of fuel cost and ignoring the fact that NPL had undertaken the washing of coal only on behalf of SECL in order to comply with the requirements of the Ash Notification, which required only such coal to be supplied and used which had Ash Content below 34%.*

(g) *Whether the State Commission erred to disallow payment of road transportation charges to NPL, while failing to appreciate that this arose due to delay in land acquisition which was a stated obligation of the State Govt., and ignoring the fact that the delay in land acquisition for a railway siding was attributable to Government of Punjab, not to PSPCL.*

(h) *Whether State Commission erred to disallow payment/recovery of energy charges based on consideration of GCV of coal on 'as-delivered to the Project' basis and as such, rejected the charges incidental to procurement of coal.*

(i) *Whether the State Commission erred to disallow payment of capacity charges to NPL for the period 20.02.2014 to 03.03.2014 when availability for the project was declared using coal procured from alternate sources, misinterpreting the Order passed by this Tribunal on 21.08.2013 and acting contrary to its own Order dated 19.02.2014 and the provisions of the PPA.*

5.3.1 Issue 1: Denial of Washing Related Costs

"We are of the considered opinion that in the absence of specific mention of washing related costs to be allowed under the cost of coal to be considered in the Energy Charge calculations, cost of washing of coal by the Appellant cannot form part of the cost of coal to the Seller for purpose of calculation of energy charges we would be constrained not to express any specific view that how a Bidder should have factored the costs related to washing of coal into their bids."

5.3.2 Issue 2: Consideration of Incorrect GCV of Coal

"As the tariff discovered under the competitive Bidding process was under Section 63 of the Electricity Act, 2003 and not under the Section 62 of the Electricity Act, 2003 based on cost plus mechanism, such interpretation to the measurement of GCV of coal as of the washed coal as received at the Project cannot be accepted".

5.3.3 Issue 3: Denial of Road Transportation Cost

"The PPA provides for signing of a separate Fuel Transportation Agreement (FTA) between the seller of electricity (NPL) and the transporting agency (Railways) besides signing of the Fuel Supply Agreement with the fuel supplier (SECL). The Commission further notes that the FTA has not been signed by NPL with railways without assigning any cogent reason. The Commission also notes that the permission for grant of surface/emergency cross-over by railways was applied very late by NPL, which, in the opinion of the Commission, should have been applied much earlier. As such, NPL should claim compensation from Railways, not from PSPCL".

5.3.4 Issue 4: Non-payment of Capacity Charges

In October 2012, NPL filed Petition No. 56 of 2012 before the State Commission owing to failure of PSPCL to fulfil its obligation of arranging sufficient quantity/quality/ grade of coal in terms of the competitive bidding guidelines/bidding documents and the PPA. The same was dismissed by the State Commission by its order dated 31.12.2012. NPL filed Appeal No. 68 of 2013 challenging the order dated 31.12.2012. This Tribunal in its Order dated 21.08.2013 passed in Appeal No. 68 of 2013 specified a mechanism for procurement of coal from alternate source to meet the expected shortfall in supply from linked source in order to operate the power plant as per the terms and conditions of the PPA. This was followed by the State Commission's order dated 19.02.2014, laying down the modalities for procurement of alternate coal. Accordingly, NPL procured alternate coal to meet the expected shortfall. During the period in question, the linkage coal was not available despite best efforts undertaken by NPL. Supply of linkage coal under the FSA commenced on 22.02.2014 i.e. within one month from the COD of Unit-1. Accordingly, NPL declared availability on the basis of coal procured from alternate sources. **The capacity charges are to be allowed during the period when use of coal from alternate sources was allowed by this Tribunal as well as State Commission based on the Availability as declared by the Appellant during that period** *(emphasis added).*

5.3.5 Issue 5: Non-payment of Third-Party Analysis Charges, Liaising Charges and Transit and Handling Losses

Being a competitive bidding process, unless the provisions of Bidding documents specifically provide for the reimbursement of any expenses or consideration of any cost in the Tariff components, it cannot be allowed after completion of the Bidding process. Bidders could have sought such queries from the Procurer before submission of the Bid. We are in agreement with the findings of the State Commission in this regard.

5.4 Supreme Court

The Supreme Court order on Nabha Power's claims, delivered on 5.10.2017 considered the arguments made by both parties and came to the conclusion that the basis of the PPA is the actual cost incurred with regard to the coal. It held that

(a) The appellants are entitled to the washing cost of coal.
(b) The transportation from the mine site via washing of coal to the project site inclusive of cost of road transportation for the period where it was necessary would be paid.
(c) The Calorific value of the coal would have to be taken at the project site.

Coming from the apex court of the land, the order would apply also to the other IPP in the state, Talwandi Sabo which had similar claims relating to coal washing.

As to the controversy on whether the cost of coal has to be determined on the basis of the purchase price from SECL at the 'mine-end', when the property is supposed to pass to the user, or whether it is the cost of coal to be used for the plant as incurred by the NPL at site of the project, or the 'project-end', the court held:

"Thus, there is no hesitation in our concluding that in view of the specific formula provided, only three aspects relatable to coal would determine the particular coefficient". These three expressions are thereafter followed by the stipulation that the coal has to be recently supplied 'to and at the project'. The question is, what is the meaning of this expression? The word 'to' obviously would have reference to transporting while the word 'at' would have relationship with unloading since it would be 'transporting to' and 'unloading'. Thus, the definition of FCOAL is the weighted average actual cost incurred by the appellant of purchasing the coal and transporting it to the project site and thereafter unloading the coal at the project site. The fact that the property in coal passed on to the appellant vis-à-vis SECL, on delivery being taken at the mine-end would not change the definition of coal pricing as is required for the purposes of calculation of the tariff.

The apex court held that in this case, the principle of 'Reddendo Singula Singulis' should be applied. Where a complex sentence has more than one subject, and more than one object, it may be the right construction to render each to each, by reading the provision distributively and applying each object to its appropriate subject. A similar principle applies to verbs and their subjects, and to other parts of speech. Washed coal is a necessity for the project as a quality requirement for the formula envisaging the requisite quality of coal to be obtained at the project site and, thus, include all the relevant costs up to that quality. The mere term 'coal', therefore, would have to mean 'washed' coal, as no other type of coal could be used in the matter at hand.

Washed Coal Cost: Coal used in a power plant has three aspects, i.e. purchase, transport and unloading all of which will determine the amount required to be reimbursed. The court applied the principle of *Reddendo* and held that delivery 'to the project' could only mean 'at the site of the project' and not at the mine site. Supreme Court also held that prior activity of washing before delivery at the project site would be part of the cost; hence, the reference to term 'coal' in Energy Charges Formula (ECF) was to, 'washed coal' and not 'unwashed coal'. The court clarified that while NPL had to arrange washing of coal, it did not imply that the cost of washing had to be borne by them. The apex court specifically applied the definition of 'business efficacy' in interpreting the ECF in a manner as would be normally understood.

GCV: Applying the principles of business efficacy as well as *Reddendo, the SC* held that GCV of the coal was to be measured at the project site where the title of the required quality of coal used for generating electricity passed to NPL.

Transportation cost: The SC observed that the transportation cost till the project site had to be compensated to NPL as it was not qualified by the methodology of transfer, i.e. rail or road. Since the railway siding had not reached the project site, as a matter of necessity, the transportation cost from point to point was required to be paid.

The SC considered the surrounding circumstances highlighted by NPL, such as representations on the basis of which the bid was submitted, bidding documents and guidelines for determination of tariff to show liability of PSPCL to reimburse additional expenses. However, since the ECF contained only three elements, the SC held that other costs incurred towards crushing, sizing of coal, etc., could not be implied.

However, even after the Supreme Court had given its verdict directing PSPCL to reimburse identified expenses of NPL, the former had not only failed to reimburse, but also disputed some of the same issues once again, compelling NPL to approach the apex court again for contempt. Talwandi Sabo, the other IPP which had claimed costs of coal washing from PSPCL had also filed a contempt petition against the latter for non-compliance with the Supreme Court order. These contempt proceedings were clubbed and decided by an order dated 07.08.2019. The court took note of the arguments which were advanced at the time of decision of the appeal(s) relating to the controversy around Schedule VII of the Power Purchase Agreement (PPA) and held that there was no reason for any doubt on the issues decided therein and ruled that its judgement be followed in letter and spirit. PSPCL then requested that the calculations based on the apex court's orders be made by the regulator, which claim was also rejected by the Supreme Court stating that the PSU was seeking to agitate the same points before the regulator after they had already been settled by itself. Concluding, the apex court warned that if any further contempt proceedings are brought before it, the consequences could be serious for PSPCL.

5.5 Analysis

When called upon to interpret the terms of a contract, both the regulator and APTEL ruled strictly by the wording of the contract to disallow claims of the power developer. However, the Supreme Court took a reasoned and fair view in interpreting the terms of the contract, as understood in a business sense, but without implying terms into it. The court pointed out that modern-day contracts are drafted by technical and legal experts with domain knowledge. Both contracting parties have access to such expertise. As such, there was no need to imply terms into the contract, but read it as it is drafted, i.e. explicitly. The apex court also stated that implied term was a concept used strictly when necessitated by the penta principle test,[4] *as elucidated later in this chapter.*

At the same time, the Supreme Court has reinforced the cardinal maxim that everyone is equal before the eyes of the law; that the State or its agencies cannot act in an arbitrary or unfair manner. The court has conclusively laid down the law that the State or its instrumentality are bound by the letter and spirit of the contract.

This judgement reinforces the sanctity of explicit terms enshrined in the contract by providing the much-needed caution for the courts to abstain from upsetting a carefully drafted contract. Where the drafting is unambiguous and the language is clear and unequivocal, the court has preferred to go by the wording rather than second-guess and insert intentions. The court is required to be mindful of contemporary technical expertise of legal drafting and, at the same time, should also not ignore the objectivity of the contract. The judgement upholds mutuality of obligations as an unalterable requirement in a contract and ensures that a party's contractual rights are based in equity.

The Court analysed international and domestic jurisprudence on the concept of implied terms and concluded that the following conditions must be satisfied to imply terms into contract, namely (1) it must be reasonable and equitable to imply terms; (2) it must be necessary to give business efficacy to the contract, so that no term will be implied if the contract is effective without it; (3) it must be so obvious that 'it goes without saying'; (4) it must be capable of clear expression; (5) it must not contradict any express term of the contract.

Yet, at the same time, where it was warranted, the court has not shied away from its duty to provide a just and fair deal to the aggrieved party. If the railway siding got delayed because of no fault of the developer, the latter cannot be penalized for performing the contract despite this difficulty by resorting to road transportation.

PSPCL's reluctance to abide by the apex court's reasoned orders can be explained but not justified. The cash-strapped discom would have had to make substantial compensatory payments to both NPL and Talwandi Sabo necessitating a steep increase in tariff to the tune of 0.35 p/kWh.[5]

---

[4] These five principles were laid out by the court based on its study of international and domestic jurisprudence.

[5] Bright spark for PSPCL as Centre removes 'washed coal' condition Hindustan Times dated 28 May 28 2020.

Curiously, GoI had recalled its earlier order mandating coal washing for all thermal power plants, giving some relief to power developers who, thanks to the judgement analysed in this chapter, would have had to bear the costs of such washing.[6]

5.6 Chronology—Nabha Power Limited

5.10.2017—Supreme Court order on NPL

2017—NPL appeals to Supreme Court against APTEL order

14.12.2016—APTEL order on NPL appeal

1.2.2016—PSERC dismisses certain items of cost in tariff; NPL appeals to APTEL

2014—Nabha Power Limited wins bid and commissions power plant and raises invoice

2009—Punjab State Electricity Board invites bids for thermal power plants.

5.7 Implying Terms into a Contract—APMuL Versus GUVNL

5.7.1 GUVNL Versus APMuL

In February 2006, GUVNL, a holding company of Gujarat discoms invited competitive bids for supply of power to the state. The bids were invited under Case 1 of GoI guidelines for competitive bidding issued in 2005 wherein the developer has the freedom to decide the location, fuel, etc., of the project. The RFP issued by GUVNL also clearly stated that the seller had to take full responsibility for tying up fuel, as the extract from the bid document give below will show.[7]

"Clause 2.2.4

The Bidder shall submit a comfort letter from a fuel supplier for fuel linkage or a proof of captive fuel sources at the time of submission of proposal in response to the RFP.

Clause 3.1.3

The size and location of the Power Station(s) and the source of fuel and technology shall be decided by the Seller. The Seller shall assume full responsibility to tie up

---

[6] Ibid.

[7] https://www.business-standard.com/article/companies/how-adani-shifted-base-and-claimed-relief-113040800012_1.html dated 8 April 2013. Accessed on 23 March 2023.

the fuel linkage and to set up the infrastructure requirements for fuel transportation and its storage.

Clause 4.1.1

...

8. Proof of Fuel Arrangement—Bidder needs to indicate the progress/proof of fuel arrangement through submission of copies of any one or more of the following

(a) Linkage Letter from fuel supplier
(b) Fuel Supply Agreement between Bidder and fuel supplier
(c) Coal Block Allocation Letter/In Principle Approval of Allocation of Captive Block from Ministry of Coal
(d) Other details submitted by the Bidder subject to being accepted by GUVNL as a sufficient proof for demonstration of ability" (*extracted from GERC order dated 31.8.2010 in Petition No.1000/2010*).

According to news reports of the time, seven bids were received, including one by an Adani Enterprises-led consortium that proposed setting up a 1200-MW domestic coal-based plant in Chhattisgarh, quoting a tariff of Rs. 2.34 per unit. This included Rs. 1.34 as energy charge and Rs. 1.00 as capacity charge—both under the 'non-escalable' head.[8] According to Adani, it had bid on the basis of commitment made by GMDC, a Gujarat government company, to supply coal from its Morga II coal block in Korba in Chhattisgarh (ibid). However, the PPA between the parties did not make performance of the contract conditional on coal allocation by GMDC.

Adani won the bid and went on to sign a Power Purchase Agreement (PPA) with GUVNL in February 2007, for supply of power to the state for 25 years. Yet, within ten days of signing the PPA, Adani went back to GUVNL, asking for a supplementary PPA to shift the project from Korba to Mundra in coastal Gujarat, without altering the terms of the PPA. In support of this request for shifting the site of the project, Adani had also submitted a MoU with imported coal suppliers, a claim that has been reiterated in the company's Annual Report (2007–08) page as well as Prospectus for the relevant years.

"The Mundra power projects are located along the coast and will utilize imported coal as primary fuel for its operations. APL has entered into long-term coal supply arrangements for importing coal with the company for Mundra power projects. PT Adani Global, subsidiary of the company, has entered into agreements with holders of long-term exploitation licenses to exclusively mine coal in Bunyu Island, Indonesia. For Mundra I, II and III power projects, APL proposes to procure the coal from PT Adani Global which will source such coal from mines in Indonesia. The coal for Mundra IV power project will also be procured internationally and supplied by the company to the power project sites. As a result of our long-term coal supply agreements and relationship with the supplier, we believe that the company will

[8] GoI guidelines allowed the bidders to bid either inflation-linked escalable tariffs or fixed or non-escalable energy tariffs which will be constant throughout the 25-year life of the project.

benefit from competitive pricing for procuring the coal for its various power projects" (page 23 of Annual Report 2007–08).

Based on these claims, on 18.4.2007, Adani Mundra Power (APMuL) and GUVNL entered into a supplementary PPA for supplying the same quantity (1000 MW) from Adani's Mundra power plant (APMuL) at the same tariff.

GMDC's assurance of coal supply for the project was never converted into a Fuel Supply Agreement even after follow-up by Adani with Government of Gujarat, the owner of GMDC. A year later, on 1.5.2008, APMuL addressed another communication to the Government of Gujarat, requesting it to impress upon GMDC to adhere to its assurance and supply the coal from the coal blocks allocated to the GMDC.

Meanwhile, in June 2008, GUVNL demanded that APMuL furnish an additional bank guarantee since it had not complied with the terms of the PPA yet. APMuL complied and deposited Rs. 25 crores with GUVNL, but also stated that its PPA with GUVNL was contingent upon getting coal supplies from GMDC, although this had not found expression in the PPA document itself. On 24.03.2009, a letter from GMDC intimated Government of Gujarat's decision to allocate to APMuL, 50% of coal from Naini coal block allocated to GMDC although the initial offer was for coal from Morga block.

Yet, no firm FSA had materialized despite reported follow-ups by APMuL. On 17.1.2009, APMuL wrote to GUVNL that it was unable to supply power as per PPA since GMDC had not signed FSA as committed. APMuL stated that, in the circumstances, it had no option but to terminate the PPA with GUVNL. On 27.02.2009, the Government of Gujarat wrote to GMDC, asking it to supply coal to APMuL from Naini block. Yet, no coal supplies were forthcoming. Parleys between the concerned parties allegedly yielded no results. Eventually, APMuL issued a termination notice to GUVNL, unilaterally terminating its PPA with effect from 4.1.2010.

5.8 Gujarat Electricity Regulatory Commission (GERC)

GUVNL petitioned GERC, contesting the unilateral termination of PPA by APMuL. The regulator declared that the PPA was not dependent on fuel supply by GMDC and that the developer was required to obtain coal from other authorized sources to supply power to GUVNL. GERC stated that APMuL could not take advantage of its own default in not procuring coal. GERC, by its judgement dated 31.08.2010 allowed the petition of GUVNL, holding that the termination of the PPA was illegal and directed APMuL to supply power to the procurer at the rate agreed upon per PPA.

5.9 APTEL

APMuL then approached the Appellate Tribunal for Electricity (APTEL) claiming that it was constrained to terminate the PPA because of non-supply of coal by GMDC. The issues before APTEL were

(1) *Whether the Power Purchase Agreement dated 02.02.2007 executed between the petitioner (GUVNL), and the respondent (APMuL) was dependent on supply of coal to the respondent by Gujarat Mineral Corporation?*
(2) *Whether the respondent (APMuL) had an obligation to arrange coal from sources other than Gujarat Mineral Corporation in order to fulfil its obligation under the Power Purchase Agreement dated 02.02.2007 with the petitioner (GUVNL)?*
(3) *What is the effect of the disclosures made by the respondent (APMuL) relating to execution of a Fuel Supply Agreement with Adani Enterprises Ltd. in its Annual Reports, Director's Report and the Prospectus filed with SEBI?*
(4) *Whether APMuL can terminate the Power Purchase Agreement dated 02.02.2007 on the basis of failure to execute a Fuel Supply Agreement and, therefore, be relieved of its obligations to supply power to the petitioner (GUVNL) under the said Power Purchase Agreement?*

On 7.9.2011, APTEL held that while the PPA allowed either party to terminate the contract if an event occurred, there must be agreement between both parties on the occurrence of such event (in this case, unavailability of coal) before either party could exercise its right to terminate the PPA. Since there was no such agreement between the parties, it held that the parties could not unilaterally terminate the PPA. In other words, "*the right of termination provided to the Seller in the PPA (Article 3.4.2) would arise when both parties agree to the happening of the certain events contemplated therein and the seller is willing to pay liquidated damages specified therein*" (APTEL 2011). APTEL also held that the option of unilaterally terminating the PPA would be available to APMuL only if GUVNL had defaulted on any of its commitments, which was not the case in this instance.

5.10 Supreme Court

APMuL then appealed to the Supreme Court which took up the matter for hearing immediately, setting up a vacation bench, something that invited an adverse remark from a senior advocate of the apex court. Mr. Dushyant Dave wrote a letter to the Chief Justice of the Supreme Court that proper court procedure was not followed in two legal cases concerning the Adani group in the Supreme Court. One of them related to APMuL.[9]

[9] https://thewire.in/law/dushyant-dave-letter-which-cases-did-adani-get-relief-in-supreme-court accessed on March 23 2023.

The main contention raised by APMuL in the Supreme Court was that the bid was submitted by it on the basis of the commitment given to it by the GMDC that it will supply the coal, first from Morga II in Korba, Chhattisgarh and then from Naini block in Odisha. Since GMDC had failed to abide by its commitment and had not executed the FSA with APMuL, there was non-compliance with the conditions stipulated in Article 3.1.2 of the PPA, and therefore, APMuL was entitled to terminate the agreement, by giving 7 days' notice in writing in accordance with the provisions of Article 3.4.2 of the PPA. So the only liability of APMuL was to pay the liquidated damages at the rate of Rs. 10 lakhs per megawatt of the contracted capacity, which worked out to Rs. 100 crores for 1000 MW.

APMuL had argued that the commission and the Appellate Tribunal had grossly erred in holding that unless there was an agreement between the parties to the effect that there was non-compliance with the conditions mentioned in Article 3.1.2 of the PPA, the appellant was not entitled to invoke the provisions of Article 3.4.2 of the PPA. APMuL further submitted that since the contract also provided for liquidated damages, the commission as well the Appellate Tribunal, ought not to have given a direction for specific performance.

GUVNL argued that the PPA, which was entered into between the parties, was not executed on the basis of commitment by the GMDC and that the procurer was not concerned with the issue as to from where APMuL would arrange for its supply of coal. "The PPA between the APMuL and GUVNL is only in respect of supply of power. It is submitted that on the GMDC's failure to adhere to its commitment to supply indigenous coal, it was the responsibility of APMuL to make arrangement for an alternative source and to enter into FSA with any other coal supplier", GUVNL stated. GUVNL also pointed out that APMuL was using imported coal in its other plant in Gujarat and could well do the same for this one. The counsel for GUVNL argued that by not making arrangements for fuel supply, APMuL had committed default and as such, could not be permitted to take advantage of its own default to terminate the PPA.

The Supreme Court invoked the principle of business efficacy on the ground that "a plain literal interpretation of the term in the agreement or the contract, it is not possible to achieve the result or the consequence intended by the parties acting as prudent businessmen. This test requires that a term can only be implied, if it is necessary to give business efficacy to the contract, to avoid such a failure of consideration that the parties cannot, as reasonable businessmen have intended. If the contract makes business sense without the term, the courts will not imply the same". The apex court decided that this dispute warranted implying terms—that the performance of the contract was contingent on GMDC supplying coal—into the contract, allowing either party to terminate the contract if it became infeasible to perform it. The court therefore allowed APMuL to terminate the contract and pay the liquidated damages to GUVNL as specified in the PPA.

"We are, therefore, of the considered view that the Appellate Tribunal has grossly erred in coming to the conclusion that Article 3.4.2 of the PPA could be invoked only in the event that there is an agreement with regard to violation of any of the conditions in Article 3.1.2. We find, that both the commission and the Appellate

Tribunal have grossly erred in arriving at finding that termination can be effected under Article 3.4.2 only if there is an agreement with regard to non-compliance of condition under Article 3.4.2 by both the parties. We are of the considered view that the Appellate Tribunal has erred both on facts and in law. We are of the considered view that the appellant was entitled in law as well as on facts to invoke Article 3.4.2 of the PPA and terminate the agreement".

Holding the termination valid and legal, the Supreme Court went on to declare that APMuL was entitled to compensatory tariff for supplying power hitherto to GUVNL by purchasing coal from the market. APMuL is entitled to "adjustment of cost of project and interest on expenditure incurred by it for completion of the Project, expenditure towards running of project after obtaining coal from open market and interest on delay of payment at the rate determined by the commission".[10] In its order dated 2.7.2019, the apex court directed CERC, the central regulator to determine the compensatory tariff payable to APMuL within three months of the order.

5.11 The Sequel

Pursuant to the judgement of the Supreme Court, APMuL filed a petition with CERC in September 2019 and eventually submitted all the data called for by CERC in February 2020. The Supreme Court extended the time given to CERC to determine compensatory tariffs. The computation made by CERC to arrive at the total compensatory tariffs payable to Adani are not available in the public realm. But according to news reports, GUVNL was required to pay APMuL, over Rs. 11,000 crores.[11]

GUVNL again approached GERC on 2.7.2019 (250/MP/2019) seeking withdrawal of its earlier order awarding compensatory tariff.

Alarmed at the prospect of having to pay such a huge sum as compensatory tariff and interest to APMuL, GUVNL approached the Supreme Court in 2019 with a curative petition against Adani, GERC and others seeking review of its earlier order on points of both law and facts. In order to file a curative petition, the petitioner has to **establish** that the **principles of natural justice were violated** and fear of the bias of the judge and judgement that adversely affected him (emphasis added). The Supreme Court had admitted the curative petition in September 2021 and constituted a five-judge bench chaired by the Chief Justice of Supreme Court. It had also issued notice to APMuL to submit its response. Curative petitions are generally decided in camera through deliberations by the bench and without new arguments by lawyers.

Curiously, on 8.2.2022, the Attorney General of India, the government's senior most law officer informed the Supreme Court Bench hearing the curative petition that the parties had settled the disputes. "*There was a claim of 10,000 crores compensation and it now stands withdrawn. So we pray for settlement to be recorded*", the AG said.

---

[10] Supreme Court order dated 2.7.2019 in APuML versus GERC.

[11] https://timesofindia.indiatimes.com/business/india-business/adani-power-gives-up-rs-11000-crore-claim-to-settle-gujarat-ppa-dispute/articleshow/88795913.cms accessed on 23 March 2023.

APMuL's counsel informed the bench that they had agreed to continue supplying power to GUVNL. The CJI asked if the settlement could be accepted in the curative plea without interfering with the judgement since it involved questions of both law and facts. But then, the apex court passed the following order: *"After we issued notice, it appears that during the pendency, parties have settled dispute & filed joined application. After hearing, the settlement deed is taken on record. In view of the of the settlement, we dispose off (sic) the curative petition saying that both parties are governed by the settlement entered into by parties. Accordingly, petition stands disposed of"*.[12]

5.12 Analysis

Supreme Court had in a dispute involving Nabha Power in 2018 held that terms could not be implied into a contract and that it would set a bad precedent. Yet, in this case which seems to have been heard and decided by a three-judge vacation bench headed by JJ Arun Mishra, the court has taken the opposite view. In fact, in a letter to JJ Ranjan Gogoi, the then CJI, Mr. Dushyant Dave, Senior Advocate of the Supreme Court had remarked that this case, along with another also pertaining to the Adani Group was heard "in complete contravention of the settled practice of the Supreme Court as also its established procedure. Both these matters were listed, taken up and heard without any justification, and in a hurry and in an improper manner. As a result, besides causing grave injury to public interest and public revenue, it has caused immense damage to the image of the Supreme Court and the administration of justice." (ibid)

It is a matter of grave concern that the Supreme Court of India, citing business efficacy considerations, should decide to imply terms into a contract where none were called for. The RFP issued by GUVNL in 2006 had made it abundantly clear to all the bidders that the bids were sought under Case 1 of GoI guidelines for competitively bid projects wherein the developer takes responsibility for tying up fuel supply. In this case, Adani Enterprises, the parent company of APMuL had also bid under 'non-escalable' tariff category quoting a levelized variable cost over the life of the project (Prayas, March 2011). Thus, Adani Enterprises knowingly took the risk of fuel contracting as well as the stability of fuel prices over the lifetime of the project. Not making the performance of the PPA contingent on coal supply was not an inadvertent error in the contract to the detriment of business efficacy as the apex court concluded, but was a conscious decision of the developer responding to the RFP that had laid down the bidding parameters ab initio. Such a conclusion is also supported by the fact that within a month of signing the original PPA to set up the project in Korba in Chhattisgarh, Adani had shifted the project location to Mundra, too far from any domestic source of coal indicating that it had planned to use imported coal.

[12] Ibid.

Besides, to imply terms into the contract on the assumption that the contracting parties would have meant to include it, but somehow omitted its mention in the PPA seems too egregious an omission and stretches credulity. Especially when coal supply agreement is at the heart of a thermal power plant, it is inconceivable that it would be excluded from the ambit of a PPA drafted by technical and legal experts.

The inexplicable decision of the winning bidder to move the project site to Mundra, a coastal location where it could import coal from its own overseas coal mines, rather than depend on GMDC's FSA raises some red flags. That this shift was sought and obtained almost immediately after the PPA was signed in February 2007 shows that perhaps Adani had intended to substitute domestic coal by imported fuel even before GMDC reneged on its assurance of FSA. Perhaps this was the reason for GMDC not converting its assurance into a Fuel Supply Agreement.

Yet, from 2007 until 2010, Adani kept up the fiction that the plant was planning to use domestic coal. It is noteworthy that the power generating equipment is specific to the type of fuel and as such, seamless transition from domestic to imported coal would not have been possible once the plant was set up. The first two units of 330 MW each had already been commissioned by 2010, presumably designed to use imported coal anyway.

This observation is corroborated by the fact that in 2011, the company petitioned CERC, seeking an increase in tariff on account of the increase in the price of imported coal. This, despite the developer's commitment under the 2007 PPA that the terms of the firms of the first PPA signed would remain unchanged even after the site shift.

This research maintains that if APMuL was aggrieved by the failure of GMDC to convert its commitment into a FSA, APMuL should have taken it up with GMDC, perhaps, even resorted to legal recourse, to force compliance with such commitment. It is also curious that GMDC failed to follow up on its commitment to convert it into a firm FSA even after prompting by its shareholder, viz. the Government of Gujarat. The liability in this case devolves on the coal supplier, namely GMDC, and not on GUVNL, to be dealt with through a separate bilateral dispute between APMuL and GMDC. Instead, APMuL chose to foist the dispute on GUVNL and the Supreme Court entertained APMuL's plea by implying non-existent terms into the PPA.

Implying terms into a commercial contract is fraught with perils. First, it is unfair to the other six bidders who lost their bid perhaps because their quotes were higher, possibly owing to realistic assumptions on coal supply. It is unfortunate that none of the bidders who lost the bid to Adani came forward to protest, intervene or petition the courts seeking fair play. Second, implying terms into a contract sets a bad precedent, inviting multiple interpretations of the contract to suit the situation. Third, violating the sanctity of contract to privilege one party results in financial burden (in the present instance, Rs. 10,000 crores) to the other contracting party and also goes against the public interest since the consumer will eventually end up paying this through power tariff or the taxpayer, through higher taxes.

In the Nabha Power case, another bench of the Supreme Court of India after meticulously analysing domestic and international jurisprudence on the concept of implied terms in contracts stated that such standardized principles should not substitute the court's own view of the presumed understanding of the commercial terms by parties. The court recognized that contracts in the contemporary era are drafted with high technical legal expertise and through a process of negotiation. The court had urged that a contract be read as per its explicit terms and has therefore cautioned commercial courts from indulging in unnecessarily implying terms into a contract and upsetting a carefully drafted contract.

In the present case, the Supreme Court has also set a precedent, perhaps unwittingly, of including a third party with no privity of contract into the application of the contract. In doing so, it might have widened the scope of 'business efficacy' principle and in the process, rendered the sanctity of contracts vulnerable to wide and contradictory interpretations in future.

In deciding to allow the parties to withdraw the curative petition in February 2022, even after finding it involved important questions of both law and facts, the Supreme Court of India has perhaps lost a valuable opportunity to re-examine the earlier decision of the three-judge bench and establish a case law to protect the sanctity of commercial contracts entered into by diligent parties. Unfortunately, the controversial judgement of 2019 might become a precedent for future decisions related to sanctity of contracts.

And finally, it is indeed strange that AMP should make a U-turn and waive its colossal claim of over Rs. 10,000 crores from GUVNL (according to the Attorney General himself) just when the Supreme Court was about to deliberate on the curative petition. Forgoing such a massive claim without any apparent reason for doing so is bound to raise unhealthy speculation.

5.13 Chronology in GUVNL Versus APMuL

2022

8 Feb: Attorney General seeks withdrawal of Curative Petition filed by GUVNL and Supreme Court allows withdrawal.

2019

2 July: Supreme Court judgement upholding implying terms into a contract.

Sep: Curative petition filed by GUVNL in the Supreme Court seeking setting aside of CERC order.

2013

- **3 Apr**: CERC order recommends 'compensatory tariff' for Adani Power.

2012

- Adani Power petitions CERC, seeking tariff increase.

2011

- **23 Sep**: Indonesian Regulation comes into effect.

2010

- **26 Jul**: Adani Enterprises enters into a consolidated coal supply agreement with Adani Power
- **31 Aug**: GERC
- **23 Sep**: Indonesian Regulation promulgated.

2008

- **5 Feb**: MoU with Kowa Company Ltd is terminated
- **18 Mar**: MoU with Coal Orbis Trading GMBH is terminated
- **24 Mar**: Adani Power executes coal supply agreement with Adani Enterprises for coal purchase.

2007

- **4 Jan**: Adani Enterprises submits its bid as part of a consortium with Vishal Exports Overseas Ltd.
- **11 Jan**: Letter of Intent issued to the Adani Enterprises-led consortium
- **2 Feb**: Adani Power signs PPA with GUVNL for supply of 1000 MW
- **12 Feb**: Adani Power proposes to supply power from Mundra project
- **18 Apr**: Accordingly, a supplementary PPA is signed between Adani Power and GUVNL.

2006

- **1 Feb**: GUVNL invites bids for power supply on a long-term basis
- **9 Sep**: Adani Enterprises signs MoU with Germany's Coal Orbis Trading GMBH
- **14 Nov**: GMDC issues letter to Adani Power for supply of coal from its Morga II block in Chhattisgarh
- **24 Nov**: GUVNL issues RFP; states fuel sourcing as the bidder's responsibility
- **21 Dec**: Adani Enterprises signs MoU with Japan's Kowa Company.

5.14 Nabha Power Ltd Versus State of Punjab—Heat Rate Degradation

Nabha Power Limited (NPL) is a supercritical thermal power plant with a capacity of 1400 MW, selected through a competitive bidding process conducted by PSPCL, the holding company of Punjab discoms in 2009 to augment the state's generation capacity. PSPCL signed a 25-year PPA with NPL in 2010 and the plant was commissioned in 2014.

In 2015, NPL petitioned the Punjab Electricity Regulatory Commission for compensation, claiming that the discoms drew much less than their committed offtake of power generated by the former, leading to degradation of heat rate and consequently, higher coal consumption for every unit of electricity generated. When a coal-based power plant operates below optimal capacity for an extended period for whatever reasons, it is forced to burn more coal (called the heat rate) to produce the same units of electricity.

NPL contended that during the time of the bid, PSPCL had stated that there was acute power shortage in the state and as such, it had committed to draw all power produced by the plant at the normative 85% PLF.[13] Under the two-part tariff policy, fixed costs are payable to the developer only if it is available at 85% capacity at all times whether or not all the available power is drawn by the discom. The developer had accordingly computed coal consumption on maximum offtake of power and had arrived at its bid price. Being the lowest cost qualifying bid, NPL won the contract and went on sign a 25-year PPA with Punjab discom.

However, after the plant was commissioned, the discom, for whatever reasons, did not draw power from NPL at 85% PLF, but much less. According to the developer, in 2014, the average plant load factor of Nabha power plant ranged between 50 and 60%, and there were also frequent fluctuations in the offtake although the plant was available to supply at 85% capacity at all times as required by the terms of the PPA. The plant built on supercritical technology required consistent high output for efficiency, but the reduced offtake by the discom led to degradation of heat rate, necessitating higher consumption of coal per unit of electricity generated by the plant.

5.15 Punjab State Electricity Regulatory Commission (PSERC)

NPL petitioned the PSERC, the Punjab regulator for compensatory tariff to cover past losses from heat rate degradation and the higher coal consumption it entailed. NPL argued that it was clearly specified in the RFP and the PPA that the present Project is to be developed on the 'Supercritical Technology'. The specifications in terms of the required temperature and pressure for setting up 'Supercritical Technology'-based plant, were clearly mentioned by the PSPCL in the RFP and PPA. Further, the PPA required the developer to demonstrate the 'Supercritical Technology' attributes during the performance test of the project. A supercritical plant is able to generate very high pressure and temperature which results into higher efficiency of such plants. As mentioned in the RFP and PPA, specific pressure and temperatures were specified which are the main attributes of a supercritical plant. The plants which operate below supercritical parameters function as subcritical plants which are known to be lesser efficient in comparison with supercritical plants. NPL argued that supercritical parameters in terms of temperature and pressure as specified in the RFP and PPA can only be achieved when a plant is operated continuously at higher PLF. In case a plant is operated at a lower PLF with a varying load factor, the high pressure

[13] In fact, every power developer is required to be available (ready) to supply 85% of its capacity at all times, and therefore, tariff is computed on this basis.

and temperature required for supercritical technology-based performance cannot be maintained and plant operates at subcritical parameters (APTEL 2018).

NPL further argued that it had acted upon the representations made by the PSPCL "qua the plant having to be developed as a 'Base Load Plant' which would be operated on 'Supercritical Parameters', the obligation to provide operating conditions to ensure operation of the Project as a supercritical plant is an implied term of the PPA" Therefore, in case of breach of such obligation by PSPCL, it is responsible to make good the losses suffered by the company. Invoking the doctrine of 'promissory estoppel' in support of its claim, NPL prayed in its petition that the discom be ordered to take 85% of power in the future. (ibid)

The discom, on the other hand, argued that under two-part tariff policy, it was anyway paying fixed costs to NPL at 85% normative availability and hence was not liable to pay compensatory tariffs for excess fuel consumed on account of low capacity utilization.

CEA Norms on Heat Rate

CEA norms for pit head thermal power plants					
Heat rate (kcal/kWh)	Sub critical				Supercritical
	Less than 100 MW	100 MW to less than 200 MW	200 MW to less than 250 MW	250 MW and above	
	2600	2600	2500	2375	2250
CEA norms for non-pit head thermal power plants					
Heat rate (kcal/kWh)	Sub critical				Supercritical
	Less than 100 MW	100 MW to less than 200 MW	200 MW to less than 250 MW	250 MW and above	
	2600	2600	2500	2375	2250

Source https://cea.nic.in/wp-content/uploads/tpp___d/2021/07/Communication_reg_ACQ_Norms_for_CEA_Website_20July21.pdf

NPL was a competitively bid project under Case 2 complying with GoI guidelines which categorically state "No adjustment shall be provided for heat rate degradation of the generating stations. Even in case of bids based on net heat rate, the bidder shall factor in site conditions, loading conditions, frequency variations etc. and no adjustment shall be allowed on the quoted net heat rate for the duration of the contract".

Therefore, the questions that came up before Punjab regulator were

(i) *Whether it is the obligation of the discom under the DPR, RFQ, RFP, PPA and the Competitive Bidding Guidelines read together to ensure that the offtake of power from the plant is consistent with a baseload plant built with the supercritical technology?*

(ii) *Whether the discom is liable to provide adequate demand ensuring operation of the plant at PLF equivalent to or higher than normative level of 85% or in the alternative, to compensate NPL for further losses due to operation at low PLF?*

(iii) *Whether NPL is entitled to be compensated for the losses suffered by it on account of operation of the Plant at low PLF of around 50–60% and the consequent heat rate degradation?*

(iv) *Whether in terms of the objectives of the Electricity Act, 2003, National Electricity Policy and the Tariff Policy, a regulatory commission is barred from exercising its regulatory power in case of generation project which has been awarded through competitive bidding?*

Punjab ERC deliberated on the dispute and came to the conclusion that NPL was not entitled to higher fuel cost compensation in this case. The regulator based its decision on the fact that the PPA contained provisions that allowed NPL to sell the power to a third party if the discom failed to procure its committed capacity. Besides, even as NPL is free to sell unclaimed power to third parties, it can still claim fixed charges at 85% availability as per the contract. The regulator concluded that the provision to sell unclaimed power to a third party had been expressly included in the PPA to forestall heat rate deterioration due to low PLF. Not having utilized this 'enabling provision', NPL cannot claim compensation for excess coal consumption, the regulator declared. The regulator also dismissed NPL's petition to direct the discom to draw power at 85% PLF in future.

5.16 APTEL

NPL appealed to APTEL against PSERC order. In 2018, APTEL upheld the order of Punjab State Electricity Regulatory Commission stating that the regulator had made convincing and cogent arguments in support of its order. It also maintained that there was no breach of contract by either party and therefore, no reason to reopen it. In its order, it said, "a PPA is a binding document which cannot be reopened or reinterpreted for consequential circumstances".

The doctrine of promissory estoppel essentially provides that if a party changes its position substantially, either by acting or forbearing from acting in a certain way, after relying upon a promise made by another party, then the first party can enforce the said promise, even in the absence of a formal contract to that effect. NPL also argued that the regulator had powers to award the compensation even when the project had come through the competitive bidding route.

In its defence, PSPCL maintained that Article 14.2 of the Competitive Bidding Guidelines expressly stated that no adjustment shall be provided for the heat rate degradation of the generating stations. It is also relevant to mention that the CEA in the 'Recommendations on Operation Norms for Thermal Power Stations Tariff Period 2014–2019' dated January 2014 (CEA Recommendations Report) has clearly dealt with the provisions in Case-II Standard Bidding Documents (SBD) and separately dealt with 'Heat Rate Degradation' factor and 'Compensation for part loading/low dispatch'.

APTEL, in its order categorically stated "We agree with the submissions of the PSPCL as well as the findings of the State Commission that once competitive bidding is concluded and PPA signed, the rights and obligations of both the parties get crystallized through PPA and it emerges to be a binding instrument for the parties...., we opine that the claim of NPL arising out of higher SHR is beyond the periphery of concluded PPA and the provisions of PPA are being scrupulously implemented by PSPCL. Hence, we do not find any rationale in re-opening or re-interpreting the provisions enshrined under the PPA".

5.17 Analysis

Nabha Power Limited's claims on compensation for heat rate degradation once again highlight the inexperience of the state-owned discoms in drafting watertight bid documents and contracts. Instead assumptions, both implicit and stated seem to underlie the positions of both the contracting parties giving rise to disputes that could have been avoided. While international jurisprudence is clear on not implying terms into a contract except when they satisfy the penta principle, there are contexts and complexities that challenge such principles. What should be the extent of flexibility in interpretation remains a moot question.

The ambiguity and the perceived possibility of flexible interpretation of the provisions of the regulations, guidelines and bid document, confer, if inadvertently, an advantage to the bidder to interpret them in its favour and thus overlook certain risks so as to quote low with a view to winning the bid.

Yet, considering it is impossible to foresee all problems that could arise in a long-term contract, the best outcome from such disputes would be to address the ambiguities in subsequent bid processes.

One salutary outcome of this dispute has been the reaffirmation of the regulator's remit in tariff determination. The apex court had already laid down the regulator's unambiguous remit in Section 63 cases, and this has been reiterated by APTEL in the present instance. According to the apex court, "regulatory power is a general one, and it is very difficult to state that when the commission adopts tariff under Section 63, it functions de-hors its general regulatory power under Section 79(1)(b). For one thing, such regulation takes place under the central government's guidelines. For another, in a situation where there are no guidelines or in a situation which is not covered by the guidelines, can it be said that the commission's power to 'regulate' tariff is completely done away with?.... The first rule of statutory interpretation is that the statute must be read as a whole. As a concomitant of that rule, it is also clear that all the discordant notes struck by the various sections must be harmonized. Considering the fact that the non-obstante clause advisedly restricts itself to Section 62, we see no good reason to put Section 79 out of the way altogether. The reason why Section 62 alone has been put out of the way is that determination of tariff can take place in one of two ways—either under Section 62, where the commission itself determines the tariff in accordance with the provisions of the Act, (after laying down the terms and conditions for determination of tariff mentioned in Section 61) or under Section 63 where the commission adopts tariff that is already determined by a transparent process of bidding. In either case, the general regulatory power of the commission under Section 79(1)(b) is the source of the power to regulate, which includes the power to determine or adopt tariff. In fact, Sections 62 and 63 deal with 'determination' of tariff, which is part of 'regulating' tariff..., Section 79(1)(b) is a wider source of power to 'regulate' tariff... in a situation where there are no guidelines framed at all or where the guidelines do not deal with a given situation that the commission's general regulatory powers under Section 79(1)(b) can then be used".

5.18 Chronology in Nabha Power Limited Heat Rate Dispute

17.5.2018—APTEL order dismissing NPL plea for compensatory tariffs

7.10.2015—PSERC rejects NPL petition for compensatory tariff for heat rate degradation. NPL appeals to APTEL against PSERC order

17.4. 2015—NPL petitions PSERC for compensatory tariffs for heat rate degradation due to lower offtake of power by Punjab discoms.

References

APTEL. 2011. Order on Appeal No.184 of 2010 dated 7th Sep 2011 in Adani Power vs GERC, GUVNL et al.

APTEL. 2018. Order on IA 1511 of 2018 in Appeal No. 241 of 2018 dated 5th Dec 2018 in GUVNL vs APMuL.

Prayas Energy Group. March 2011. Prayas discussion paper on capacity addition through competitive bidding in India private sector participation in generation: From MoUs to competitive bidding.

Chapter 6
Market-Making in Electric Supply

Section 6.1 introduces the provisions in Electricity Act 2003 relating to competition in the distribution segment of electric supply industry. Section 6.2 traces the implementation of retail competition in Mumbai as it unfolded on the ground and the confusion and disputes arising therefrom; in Sects. 6.3 to 6.6, we discuss the perspectives and interpretations of various stakeholders—licensees, consumers government, the regulator, Appellate Tribunal for Electricity (APTEL) and finally, the Supreme Court of India on the interpretation of the provisions of the law on retail competition and how they shaped the outcomes in Mumbai thus far. In the process, we highlight the convergences and divergences in the perspectives of various stakeholders. Section 6.7 summarizes the findings and conclusions of our research and presents our analysis as to why regulatory fiat is insufficient to operationalize retail market competition which is fraught with difficulties not least, on account of the specificities of the Indian context.

6.1 Introduction

Neoliberal economic reforms have been premised on the rationale that competitive markets deliver the best economic outcomes. Yet, historically, monopoly market has been the default structure for vertically integrated network utilities like electric supply. This is because, utility infrastructure entails universal or near-universal coverage and as such, requires extensive network infrastructure, with attendant economies of scale which favour a single player. Network utilities are, therefore, referred to as 'natural' monopolies.

S. Mahalingam and K. Mahalingam, *Independent Energy Regulation in a Developing Economy*, https://doi.org/10.1007/978-981-99-4398-2_6

Thanks to a wave of neoliberal thinking which privilege the primacy of markets over all other governance structures, the Indian government had, in 1991, initiated structural reforms in the economy. These reforms, sought to introduce competition in sectors hitherto operating as monopolies. The power sector was one of the earliest to be reformed, opening up power generation to private investors. This was followed by unbundling of vertically integrated electric utilities in the late 1990s. In 2003, parliament passed a comprehensive law—Electricity Act 2003 which, inter alia, sought to introduce competitive markets in electricity distribution. The preamble to the act sets out the following objectives: "An act to consolidate the laws relating to generation, transmission, distribution, trading and use of electricity and generally for taking measures conducive to development of electricity industry, promoting competition therein, protecting the interest of consumers and supply electricity to all areas, rationalization of electricity tariff, ensuring transparent policies regarding subsidies, promotion of efficient and environmentally benign policies….."

Electricity Act 2003 has two explicit and distinct provisions to facilitate contested markets in power distribution. These are not exhaustive, but significant first steps. First, Proviso 6 of Section 14 of Electricity Act 2003 allows a parallel distribution licensee in an area already serviced by an incumbent licensee to build its own network to supply any consumer desirous of migrating to the new provider, subject to certain conditions. The proviso reads "Provided also that the Appropriate Commission may grant a licence to two or more persons for distribution of electricity through their own distribution system within the same area subject to the other conditions or requirements under this act comply with the additional requirements as may be prescribed by the central government and that no such applicant, who complies with all the requirements for the grant of licence shall be refused grant of licence on the ground there already exists a licensee in the same area for the same purpose". The additional requirements relate to capital adequacy, creditworthiness, etc.

Second, Section 42 contains a provision enabling 'open access' wherein the incumbent discom is required to provide non-discriminatory open access to its wires to both generators and consumers contracting directly with each other. "The State Commission shall introduce open access in such phases and subject to such conditions (including cross-subsidies, and other operational constraints)". The proviso adds that in five years, the State Commission shall provide open access to all consumers who consume more than one megawatt. This provision acknowledges the monopoly character of supply infrastructure and yet seeks to enable competition in retail supply, by allowing competing suppliers to use the incumbent's network to supply to their customers, albeit after payment of certain levies that include wheeling charges, banking charges, standby charges, cross-subsidy surcharge, etc. In industry parlance, this is referred to as separation of carriage from content.

6.2 The Sisyphean Journey Towards Consumer Choice

Mumbai is India's largest city and one of the earliest to be electrified. Traditionally, Mumbai has been served by four electricity companies—BEST (Brihanmumbai Electric Supply and Transport Undertaking) a transport-cum-electric utility run by the local government, Tata Power Company (TPC) a generator and bulk supplier to other distribution licensees. BSES (Bombay Suburban Electric Service, which later became Reliance Energy Limited and then RInfra) and the state-owned MSEB (Maharashtra State Electricity Board) whose successor entity later became MSDCL. There is an overlap of jurisdiction between BEST and TPC and again, between RInfra (REL) and TPC.[1] BEST had been licensed in 1905 and TPC, in 1907, both under IE Act 1903. TPC had begun own generation of power from 1915. BSES had taken over the distribution licence issued originally to Killick Nixon in 1926 (Supreme Court, May 2009). Since 1915, TPC had been bulk supplier to both BSES (which later became RInfra) and BEST both of which were engaged in retail distribution in different parts of suburban Mumbai. But its license issued in 1907 also allowed it to supply bulk consumers consuming over 1000 kVA, located in the jurisdiction of BSES as well as BEST.

In 1995, when RInfra set up its own generation of a 500 MW plant in Dahanu, it began to reduce its offtake of bulk power from TPC, forcing the latter to look for other consumers to supply the quantum thus rendered surplus. TPC found consumers within RInfra's license area and started supplying them directly. Over the next few years, TPC succeeded in luring some consumers away from RInfra, owing to the lower tariffs it offered, although the actual migration was not significant at this stage (Deo 2021).[2] Since TPC was building its own network to service consumers, it attracted none of the levies applicable in an open access regime.

Meanwhile, Maharashtra ERC had been set up in 1999 under the ERC Act 1998 and had begun regulating the sector. In July 2002, Reliance complained to the regulator of violation by TPC of the terms of licence issued to it in as much as the latter extended its supply to non-bulk consumers (case 14 of 2002). MERC, the regulator did not find any infirmity in TPC's extant licence to prevent it from supplying power directly to retail consumers. In fact, MERC decided that Indian Electricity Act 1910 and Electric Supply Act 1948 did not contain any provisions prohibiting competing discoms in the same licence area.

This was prior to the coming into force of EA 2003. But even as competition was unfolding organically in Mumbai's retail market, EA 2003 with its provisions for parallel licensing and 'open access' came into force. Since incipient competition

---

[1] Initially known as BSES, in 2003 this business was acquired by India's largest private sector conglomerate Reliance Industries. Following the takeover, it was renamed as REL and subsequently, renamed as RInfra from 2008. In 2018, the Adani Group purchased the distribution business from Reliance (specifically, the Anil Dhirubhai Ambani Reliance faction which had control of the business following a much-publicized family division of the assets between the two Ambani brothers).

[2] https://www.financialexpress.com/opinion/delicensing-distribution-lessons-from-mumbai/2191438/ accessed on 23 March 2023.

in retail supply had already started in Mumbai, MERC decided to give teeth to the parallel licensing provisions in the act, but was caught in a bind since TPC's licence conditions did not prohibit retail supply, and it had already established a 'fait accompli'. However, vide the same order, MERC restricted TPC from offering new connections below 1000 kVA in the licensed area common to both RInfra and TPC, on the ground that there was no level playing field between TPC and RInfra.

TPC, not hamstrung by regulated tariff (in the initial stages), was able to lure consumers away from the incumbent by offering attractive low tariffs.[3] To add to this anomaly, if the new entrant is willing to set up its own distribution network (duplicating the existing infrastructure) rather than availing 'open access' to the incumbent's network, it is not even hamstrung by cross-subsidies or other levies applicable to suppliers using the incumbent's networks. And if the new entrant has its own generation, as TPC did, it could offer attractive discounts to beat down competition.

As expected, TPC targeted mostly large consumers paying high, cross-subsidizing tariff (referred to as cherry picking) rather than the cross-subsidized category.[4] For the large consumers with substantial electricity bills, it became beneficial to switch supplies to TPC which could offer low tariffs without the burden of cross-subsidy. Even with the cost of new infrastructure which had to be laid to service the migrating consumers, TPC's tariffs were low enough to lure the consumers away from RInfra (BSES).

Over the years, the trickle of migration from RInfra (BSES) to TPC began to swell considerably. With TPC offering 13–35% lower tariff to Commercial LT consumes and 37% savings to HT industrial consumers, many cross-subsidizing consumers of RInfra migrated to TPC, leaving the former to service the cross-subsidized category of consumers with a lower number of cross-subsidizing consumer base (Prayas 2017).

[3] Only the initial tariff of the new entrant is exempt from regulatory purview. Subsequent tariff increases of the new entrant had to be approved by the regulator. When there is more than one distribution licensee, the regulator computes the average revenue requirement of each licensee separately and apportions it among its respective consumer categories. If the costs of one licensee are lower than that of its competitor, its tariffs would still be lower.

[4] In India, average tariff is distributed among various consumer categories in such a way that certain consumer categories cross-subsidize others. For instance, industrial and commercial tariffs are way higher than the average tariff while domestic consumers pay lower tariffs. Agricultural supply is free in many states and is priced very low in others. Thus, there is a significant cross-subsidy element in the tariff to enable the discom to cover its cost of supply.

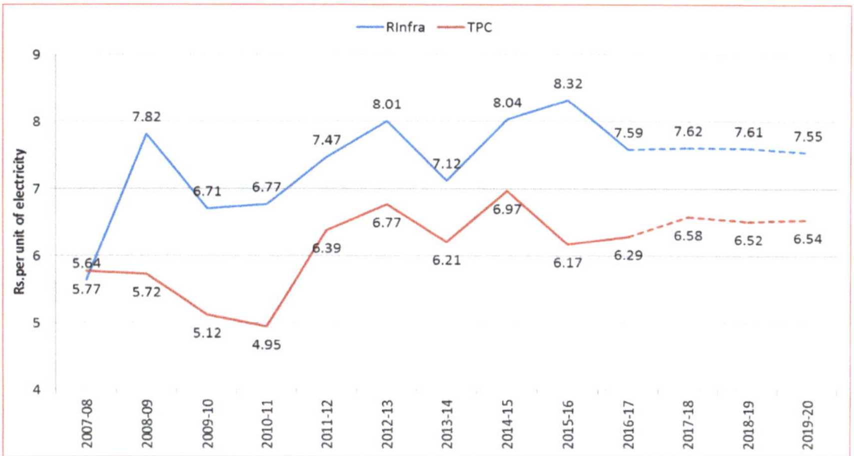

Difference in the average cost of supply of RInfra and TPC. *Source* Prayas Energy Group

Not only did RInfra (BSES) lose its large cross-subsidizing consumers, but the migrating consumers also escaped the burden of regulatory asset charges when they left RInfra (BSES) network (see Box below).

Regulatory Asset

An anomaly not originally envisaged in the tariff design but evolved over the years in response to contingent situations is 'regulatory asset'. As part of the reform package, after the establishment of independent regulatory commissions mandated to fix cost-reflective tariffs, the respective governments had withdrawn budgetary support to discoms, leaving them to fend for themselves. Section 43 of Electricity Act 2003 mandates the state government to provide budgetary subvention if it desired to subsidise a particular category of consumers (for political or other reasons), but often, state governments do not have the resources to fund direct subsidies. The hapless regulator is left to balance the interests of various consumers through cross-subsidies. Yet, as the cost of supply goes up, the scope for cross-subsidies shrinks progressively and regulators find it difficult to shift the entire cost to supply to the tariff through cross-subsidization to avoid tariff shocks in a particular year. So they hit upon an ingenious idea of setting aside costs which could not be recovered through tariffs. Termed 'regulatory asset', the idea is that these would be passed on to the tariff in subsequent years, but as the years go by, regulatory asset becomes too huge to be converted into tariff. Meanwhile, not earning cost-reflective average tariffs (regulatory asset is a notional figure and cannot even be leveraged to raise debt) seriously impacts the financial viability of the discoms. The table below gives a snapshot of regulatory assets held by a cross-section of discoms.

Regulatory Assets as of 31 March 2021 (Rs. Crore)	
BESCOM	5915
CHESCOM	139
GESCOM	266
HESCOM	0
MESCOM	927
Karnataka	**7247**
MSEDCL	19,079
Maharashtra	**19,079**
WBSEDCL	19,581
West Bengal	**19,581**
State Sector Total	**45,907**
BRPL	10,892
BYPL	9503
TPDDL	5512
Delhi	**25,906**
AEML	439
Maharashtra	**439**
NPCL	−21
Uttar Pradesh	**−21**
CESC	3233
IPCL	79
West Bengal	**3312**
Private Sector Total	**29,637**
Grand Total	**75,543**

Source https://www.pfcindia.com/Home/VS/29 accessed on 23 March 2023

Inevitably, this led to disputes not only between the licensees, but also from remaining consumers of the incumbent who now will have to bear regulatory asset charge disproportionate to their own consumption to make up for the loss from consumers who migrated to the other supplier.

These disputes, first adjudicated by MERC, were escalated to APTEL and ended up in the Supreme Court of India. In the following pages, we trace the course of these disputes and how they were decided by the three adjudicatory forums—the regulator, APTEL and the Supreme Court of India. In doing so, we highlight the differing perspectives of the stakeholders—licensees, consumers, the regulator and the courts—on the provisions that envisage retail competition in EA 2003.

6.3 Regulatory Missteps

MERC was established in 1999 by Maharashtra state government under Section 14 of the Electricity Regulatory Commission Act, 1998 (14 of 1998). When this act was replaced by the Electricity Act, 2003, State Electricity Regulatory Commissions established under Section 14 of the Electricity Regulatory Commission Act, 1998, were allowed to continue under the new legal dispensation.

When TPC began poaching RInfra consumers around 1995, the latter started offering discounts and rebates in tariff to its customers to prevent their migration to TPC. RInfra also stopped buying bulk power from TPC, as it had been doing till then, leaving TPC with surplus power for which it had to find new consumers. With its price advantage eroding because of RInfra's discounts and rebates in tariff, TPC had approached MRTPC, the monopolies regulator, alleging unfair price competition and restrictive trade practices by RInfra. In its order dated 28.8.2002, MRTPC refrained from expressing any opinion as to whether or not the alleged action of offering rebates by RInfra constituted restrictive trade practice or not. Instead, it ordered TPC to approach MERC, the sectoral regulator (MRTPC 2002).

In the event, in 2003, both RInfra and TPC were before the regulator, seeking resolution of their dispute. While RInfra claimed violation of licence conditions by TPC, TPC refuted that claim and pointed out that it was left with no alternative but to find new consumers for power rendered surplus by RInfra's reduced offtake of bulk power from itself. It claimed it had to diversify its consumer base to improve utilization of its assets.

In a petition filed before the regulator on 16.1.2003, TPC prayed that RInfra should

recover from its consumers the difference between the tariff charged by RInfra consequent on TPC's entry into its service area and the tariff fixed by the commission

forthwith cease and desist for all times, the practice of manipulating its tariff, either by grant to any consumer, rebate or discount in tariff or any other manner whatsoever and

also to direct RInfra to recover from the consumers the entire amount that was 'illegally' offered as rebate, either through its periodical bills or otherwise by withdrawing earlier bills and preferring amended bills at authorized tariff rates.

TPC also prayed for a direction to restrain RInfra from granting any rebate or discount to its consumers in any form pending disposal of its Petition and also to direct RInfra to initiate recovery from all customers the amount of rebate/discount paid by RInfra.

In this context, the questions that came up before the regulator were

Whether TPC's existing licence issued under the Electricity Act 1910 and ESA 1948 allowed it to enter retail supply as a parallel licensee

Whether the rebates given by RInfra to retain its consumer base should be recovered through supplemental tariffs as prayed for by TPC which wanted to retain its price-edge.

In deliberating these questions, the commission was guided by provisions of Section 22(2)(e) of the commission's power of regulating and promoting the working of licensees in an efficient, economical and equitable manner and, in general, against the provisions of Section 22(1)(d) enjoining the commission to promote competition, efficiency and economy in the activities of the electricity industry under ERC 1998. EA 2003 had not come into force then.

After considering the arguments put forward by the litigants and the provisions of the licenses and statutes, the commission came to the conclusion that the license to supply energy "for all purposes including supply to other licensees for their own purposes and in bulk" gave TPC an unfettered right to supply energy directly to all or any consumers in the RInfra area of supply *but no obligation to supply power* (emphasis added) MERC further held that strictly speaking, there was nothing in either the 1910 Act or the 1948 Act which bars the grant of licences to more than one party for similar purposes in the same area, which, in effect, is also the aim and object of the then just enacted EA 2003, which replaced all the earlier electricity laws.

The commission held that the rebates or discounts constitute a departure from the pre-existing tariff without the commission's approval. It further held that the burden of past rebates cannot be recovered through supplemental tariffs from consumers, but shall be borne entirely by RInfra (BSES).

MERC ultimately disposed of the petition filed by RInfra (BSES) with the following directions:

(i) *TPC and BSES should file the terms of reference for engaging a consultancy firm to study the issues relating to Section 14 of the Electricity Act 2003.*
(ii) *Select a consultancy firm/s (if need be, international level firms may be considered for selection, severally or jointly with Indian firms) for the purpose;*
(iii) *The cost of the study should be equally shared by both parties.*
(iv) *The study report should be widely disseminated among stakeholders in the city.*
(v) *Adoption of the report would be decided after a public hearing; and Implementation of the report would be undertaken as per the Commission's Regulations.*

While disposing of the petition MERC also restrained TPC from offering new connections to new consumers with energy requirement below 1000 kVA on the basis of an agreement purported to have been arrived at between TPC and RInfra (BSES) that they would maintain the status quo till the disposal of the petition and would not encourage any existing consumer to shift from one to the other (Prayas 2017).

6.4 APTEL's Volte-Face

Appellate Tribunal for Electricity (APTEL) was established in 2004 under the provisions of EA 2003 to decide on appeals against the quasi-judicial orders of regulatory commissions set up under the same act.

The decision of MERC in allowing TPC to engage in retail distribution in Mumbai was appealed by both RInfra and TPC. RInfra questioned the findings of MERC that under the licences issued to TPC it could also undertake retail supply of energy directly to retail consumers and prayed for setting it aside. TPC also questioned the order passed by MERC restraining it from offering new connections to consumers with a maximum demand of less than 1000 KVA and also to stay the directions of MERC directing the parties to engage a consultancy firm to study the issues.

The issues before APTEL were

(i) *Whether grant of rebate by RInfra (BSES) to its consumers is illegal and violative of statutory provisions of the Central Acts (54 of 1948, 9 of 1910 and 14 of 1998) and Tariff Notification?*
(ii) *Whether TPC is an aggrieved party? Whether it could maintain a complaint? Whether such complaint is bona fide?*
(iii) *Whether the directions issued by MERC are sustainable or liable to be set aside?*
(iv) *Whether the Order of MERC is liable to be interfered?*

Both the appeals were disposed of by APTEL by a common judgement dated 22.5.2006, inter alia, holding that Tata Power had *not* been granted licence to undertake retail distribution of electricity in the area within which REL had been distributing power in retail to customers directly (emphasis added). The order and findings recorded by MERC in this regard were set aside. It was also held that Tata Power could undertake only bulk supplies to licensees such as RInfra (BSES) under the licenses held by it.

On the question of rebate, the tribunal decreed "the grant of rebate by REL to its consumers is neither illegal nor it is violative of statutory provisions" of extant laws. "Being a licensee to supply power in bulk and being a generator, TATA Power cannot be termed as an aggrieved party in so far as REL giving rebate to its consumers. At the best (sic) payment of rebate may be termed as unhealthy competition resorted to between the two Houses, who are engaged in generation and supply of energy in the city of Mumbai". (ATE, 22.5.2006)

6.5 The Supreme Court Opens up a Labyrinth

It was inevitable that this dispute would be escalated to the highest court of the land. Both parties approached the Supreme Court. RInfra (BSES) complained of alleged encroachment by TPC within its area of supply and argued that the right to such open access was available to the consumer and not to the distribution licensee which had to cater to the demand made by the consumer. Two appeals were filed by Tata Power Company Limited and M/s MIDC (Marol Industries Association), against the judgement of the Appellate Tribunal. Though MIDC was not a party before the MERC, it had been allowed to intervene as a party affected by the orders of the Appellate Tribunal.

RInfra's prayers before the Supreme Court were as follows:

(a) *That TPC be restrained from in any manner selling, supplying and distributing electricity to the consumers situated within the area of supply of RInfra in contravention of the terms and conditions of their licenses and the policy of the Government of Maharashtra.*

(b) *That TPC be ordered to pay RInfra (BSES) or to the Government of Maharashtra all profits and gains made from January 1998 until TPC discontinued sale of energy to such consumers, i.e. situated in RInfra's licensed area of supply and restrained from poaching its consumers with a requirement of less than 1000 kWh.*[5]

TPC contested the restraint imposed by MERC on supply to consumers requiring lower than 1000 kVA. It argued from the 1903 Act it had always been the intention of the government to introduce competition in the electricity trade. Marol Industries argued that even if two interpretations of the law were possible, the one favourable to the consumer should be accepted by the court, viz, allowing competing suppliers.

In a landmark judgement delivered on 8.7.2008, the Supreme Court under JJ Altmas Kabir came to the conclusion that in dealing with the appeals filed both by RInfra and TPC, the Appellate Tribunal for Electricity misinterpreted the provisions of the licences granted to TPC for supply and distribution of electrical energy. The apex court further observed that the very fact that wheeling and open access had been introduced in EA 2003 showed that even a second licensee without own network could engage in retail distribution. While conceding that the submissions made by TPC counsel on open access were irrelevant to the consideration of the issue being adjudicated, the court, nevertheless, observed, "It is no doubt true that Section 42 empowers the State Commission to introduce a system of open access within one year of the appointed date fixed by it and in specifying the extent of open access in successive phases and in determining the charges for wheeling having due regard to the relevant factors, but the introduction of the very concept of wheeling is against Mr. Venugopal's (counsel for RInfra) submission that not having a distribution line in place, disentitles TPC to supply electricity in retail directly to consumers even if their maximum demand was below 1000 kVA. The concept of wheeling has been introduced in the 2003 Act to enable distribution licensees who are yet to instal their distribution line to supply electricity directly to retail consumers, subject to payment of surcharge in addition to the charges for wheeling as the State Commission may determine".

The apex court added "the provisions of both the 1903 and 1910 Electricity Acts encourage competition in the electricity trade and the same is also incorporated in the licences issued in favour of the distribution licensees, which also include licensees generating power for supply. The element of competition has been included in the Preamble to the 2003 Act and permeates the same in its various provisions... the act is meant to be consumer-friendly and one of the objectives it sets out to achieve is to give the consumer an option to choose the distribution licensee from whom it wishes

---

[5] For the sake of simplicity, the other details not relevant to our analysis have been left out.

to receive supply of electrical energy. The intervention of MIDC, Marol Industries Association and the appeals filed by it, has obviously been made in that context".

In conclusion, the apex court said: "We quash the orders passed both by MERC and the Appellate Tribunal for Electricity and allow all these three appeals upon holding that under the terms and conditions of the licences held by it, Tata Power Company Ltd. is entitled to effect supply of electrical energy in retail directly to consumers, whose maximum demand is less than 1000 kVA, apart from its entitlement to supply energy to other licensees for their own purposes and in bulk, within its area of supply as stipulated in its licences…".

This observation of the Supreme Court in fact became the foundation for operationalizing retail supply competition in Mumbai, although it still contains many ambiguities and infirmities leaving it open to more than one interpretation (Prayas 2017).

6.6 The Sequel

6.6.1 Prometheus Unbound

Acting on the Supreme Court order of 8.7.2008, MERC announced an interim changeover protocol on 15.10.2009 (Case No 50 of 2009) wherein it approved the second licensee TPC to use the lines of the incumbent, viz. RInfra to service the consumers migrating from the latter to the former. It must be noted that TPC had originally entered the licence area as a parallel licensee, extending its own lines to migrating consumers. Now, however, consequent to the Supreme Court judgement, the regulator introduced 'open access' to allow TPC to use RInfra lines, but without simultaneously determining the cross-subsidy surcharge to compensate RInfra for loss of revenue from migrating cross-subsidizing consumers. Additionally, both TPC and RInfra were given the option to fulfil their respective Universal Service Obligations (USO) under Section 43 of the EA 2003 by using the existing network of the other licensee. USO refers to the obligation of the licensee to extend connectivity and supply every consumer on demand.

Naturally, TPC's tariffs were lower, also because they had no cross-subsidy component and large number of consumers chose to migrate to TPC. According to figures furnished by RInfra to MERC, as many as 2.21 lakh out of the 28 lakh consumers of RInfra chose to move to TPC for their supply. These included a few large consumers including Mumbai International Airport Limited (MIAL), HDFC Bank, etc. RInfra found that the share of its cross-subsidizing consumers dropped from 56 to 41% consequent to this migration (MERC, Aug 22, 2012). In the next three years, nearly 85% of migrating volume was from industrial and commercial consumers while only 12% constituted the subsidized category, lending credence to fears of cherry picking on account of open access.

Alarmed, RInfra petitioned MERC on 8.12.2011 alleging that Tata Power was indulging in cherry picking, leading to a skewed consumer mix for RInfra. It further claimed that if this trend continued, the tariff of subsidized consumers who now constitute the majority of its consumer base will have to increase exponentially, eventually driving them also to migrate to TPC's supply. It prayed that MERC modify its order to disallow TPC from using its network and build its own; that it would be necessary to restore level playing field between the licensees.

MIAL, a large commercial consumer and a consumer organization (Bharatiya Udhami Avam Upbhokta Sangh) too intervened in the petition to insist on their right to choose their supplier. Both claimed 'open access' avoided wastage of resources and as such, should be allowed to continue.

MERC, after scrutiny of records submitted by the two discoms arrived at the conclusion that a very high number of subsidizing consumers were indeed changing over to take supply from TPC. On 22.8.2012, MERC ordered that prospectively, changeover will be allowed from RInfra (BSES) to TPC only for the residential consumers who consume electricity up to 300 units a month, a reversal of its earlier position. This would be reviewed after a year. RInfra (BSES) was ordered to roll out its network within its entire distribution area in the medium term within a time frame of two to three years so that TPC would be in a position to supply any consumer in its area within a minimum period of one month as required under Section 43(1) of Electricity Act 2003. On RInfra's prayer for cross-subsidy surcharge and regulatory asset charge, MERC stated it would require more time to study the matter and therefore would issue a separate order.

TPC appealed against this in APTEL and made MERC a respondent in addition to RInfra (BSES). TPC's primary grievance was about limiting competition to small consumers. RInfra (BSES), in its defence, alleged that TPC was laying its own network to service large consumers, but was using open access to serve small residential consumers. It prayed for disallowing any open access to its network.

APTEL, in its order dated 28.5.2020, concurred with MERC's finding that there was not enough evidence to maintain that TPC was cherry picking when it did not lay down LT network to service its migrating residential consumers who continued to use RInfra network. It also ordered that TPC should roll out a parallel network only in areas where it will improve reliability of supply. Finally, APTEL upheld the consumers' right to choice of supplier.

APTEL held, "where a reliable distribution system of RInfra is already existing and physical constraints in laying down of network by Tata Power and very high cost involved in the same, it is in the overall interest of consumers of Tata Power and RInfra that the changeover consumers continue to get supply from Tata Power on the RInfra's network. It will also be convenient and economical for the consumer to changeover back to RInfra in case RInfra's tariff becomes more attractive in future.... Laying down of Parallel network in a congested metropolitan city like Mumbai where a reliable distribution network is already existing is to be viewed differently from situation in other areas in the country where there are deficiencies in the existing distribution network resulting in constraints in maintaining a reliable supply to the existing consumers and extending supply to new consumers. Practical difficulties in laying down the network and extending the 11/0.4 kV network all around the congested areas in multi-storeyed buildings and narrow lanes of slums and the extremely high cost involved in making an unnecessary expenditure has to be considered. In some areas it may be practically impossible to lay down the Parallel network by Tata Power due to space constraints. Tata Power itself has stated that it is facing practical difficulties to lay down the distribution network".

However, APTEL also maintained that TPC cannot pick and choose customers and lay down distribution network selectively, but extend its lines only to improve reliability of service to the consumer. If that required an amendment to TPC's licence, it should be taken up by the regulator through due process of law. Where networks have been laid already, consumers will continue to procure their supply from their chosen supplier and switchover to the other when they please, in accordance with protocols which the regulator would draw up.

APTEL decreed that building new infrastructure will henceforth be vetted by a new institutional mechanism—Mumbai Network Distribution Assessment Committee—to assess the cost-effectiveness of the proposals of the licensees releasing supply to applicants in areas where neither licensee is present and to verify the claims of the licensees in other areas.

6.7 Summary: Untangling Retail Competition

The story of competition in Mumbai's retail electric supply, as it has unfolded over the past 12 years, has been one of regulatory uncertainty caused by trial and error. Hasty and ill-considered measures characterised regulatory decisons that failed to foresee the consequences of introducing a parallel licensee in a traditionally monopoly service area, that too in a crowded city with extensive pre-existing coverage. Lack of conceptual clarity led to avoidable confusion, not to mention the tangle of protracted, vexatious and costly litigation. Initially unregulated and subsequently partially regulated in response to emerging challenges, contestation in Mumbai's retail distribution led to wastage of resources without serving the interests of the consumers, as the law envisaged.

An analysis of the provisions of the law, the actions of licensees and consumers and the orders of the regulator, APTEL and the Supreme Court reveals that despite the tortuous process through which contestation was conducted and despite judicial intervention at the highest level, there is no clarity on whether parallel licensing is warranted at all, and if it is, when and how it should be implemented. Even at the time of writing, wringing out the benefits of contestation for the consumers without upsetting security of supply continues to remain a challenge.[6] As Pramod

[6] Security of supply entails a fallback supplier at all times, to continue supply when an existing supplier exits the market for whatever reasons.

Deo, the earlier Chairman of MERC put it, retail competition in Mumbai is still 'work-in-progress'.[7]

In this section, we analyse the contribution of each of the stakeholders in shaping the outcomes in retail competition in Mumbai's electric supply industry.

(i) **The Legal Framework: Premature and Inappropriate**

The root cause of the regulatory and legal tangles that afflict Mumbai's retail electricity markets can be traced to the ill-conceived provisions of the law which seek to prematurely introduce competition in a natural monopoly industry. Additionally, there are ambiguities and infirmities in the law itself and in the licences issued under the laws, which make them amenable to differing interpretations by different stakeholders. While no law can address all potential eventualities, a basic minimum clarity ex-ante on key issues such as cross-subsidy charges and regulatory assets is imperative to realize the anticipated gains from contestation.

Parallel Licensing

IEA 1910 *Section 3 (2) (e)* states: The grant of a license under this part for any purpose shall not in any way hinder or restrict the grant of license to another person within the same area of supply for a like purpose.

While this clause technically allows a parallel licensee in a single distribution area, it must be remembered that it was enacted in the year 1910 when electric supply was still a novelty in India and electricity networks were sketchy, largely confined to industrial and occasional urban conglomerations where it was used mainly for street lighting. In such a scenario, encouraging parallel licensing could lead to better coverage of areas in need of electric supply.

But by the time EA 2003 came into being, almost a century later, the situation had changed drastically with near-universal coverage of networks in most urban areas and large parts of the hinterland as well. Yet, EA 2003 went ahead and greenlighted parallel licensing.

EA 2003 envisages competition in distribution through two routes—open access and parallel licensing, in furtherance of consumer interest. The basic premise underlying both these provisions is electricity is just a commodity to be supplied by markets to those who can pay for it. Yet, in the modern world, electricity fulfils many roles—satisfying basic human need for lighting, heating and pumping water in high-rise apartments, etc., without which a life of dignity is impossible, as well as meeting luxury demand for air conditioners, heated swimming pools, etc. It is also an essential input into industry and commerce. Governments worldwide include affordable electric supply in their welfare schemes, so that basic human needs are met regardless of the consumers' ability to pay. Acknowledging its indispensability, governments in some cold countries forbid their utilities to switch off electricity or gas used for heating purposes even when the consumer fails to pay for it. Many developing countries including India have adopted differential pricing for different categories

[7] https://www.financialexpress.com/opinion/delicensing-distribution-lessons-from-mumbai/2191438/ accessed on 24 March 2023.

of consumption with some cross-subsidizing the others. For instance, agriculture is supplied at zero tariff in some states and at very low tariff that does not cover the cost of supply; domestic supply is priced below cost whereas industry and commerce pay steep tariffs that cross-subsidize domestic and agricultural consumption. That being the case, treating electricity like any other commodity to be supplied by competitive markets is problematic because markets tend to price commodities at marginal cost to serve each category of consumers. This would have meant lower tariffs to high-tension and extra-high tension consumers (industry) where line losses are least and very steep tariffs to rural consumers and agricultural supply which entail long transmission lines.

Mandating competition through parallel licensing entails duplication of infrastructure which will eventually be paid for by the consumers. This would lead to mergers and shakeouts, reducing the number of players to two or just one, as Joskow points out.

While Section 42 makes it mandatory for the licensee to supply everyone on demand (universal supply obligation), there is little clarity on who will be the supplier of last resort in the event one of the licensees decides to exit the business for whatever reasons. A competitive market has distinct rules of entry and exit which are missing from EA 2003. In the case of an essential service like electricity, exit of an incumbent supplier is not really an option unless there is a fallback supplier.

Therefore, Article 14 (proviso 6) which allows a parallel licensee in a distribution area appears to be an ill-conceived provision with scant regard for wasteful duplication of infrastructure which it authorizes in a poor country like India. It is hardly surprising that this provision created a complex web of cascading litigation wasting the time of regulators as well as the higher judiciary and the resources of stakeholders to no one's benefit. It is time the government wakes up to this reality and introduces an amendment to repeal this unfortunate provision of the law.

Open Access: Open access through separation of carriage from content or wires from electrons has been promoted by World Bank in jurisdictions where it oversaw reforms in the industry.[8] The outcome of separation of carriage from content has not been an unmixed blessing in the UK and EU where small (domestic) consumers are yet to see the benefits of competition. This, despite the fact that unlike Indian discoms saddled with long-term Power Purchase Agreements, the retail supply utilities in the UK and EU have moved to a real time or day-ahead wholesale market with many competing suppliers offering competitively priced supply. Even so, after the initial reduction, electricity prices have been rising sharply in recent years in the UK and EU, and the promised benefits of lower prices due to competitive retail markets have remained illusory.

In India, the ground realities are such that it is difficult to wring out the benefits of competition to electricity consumers unless all suppliers have their own generation or

[8] https://documents1.worldbank.org/curated/en/815021468042283537/pdf/More-power-to-India-the-challenge-of-electricity-distribution.pdf page 137 (accessed on 23 March 2023).

can buy all their power from real-time day-ahead wholesale markets. For many government discoms, own generation comprises a small proportion of total power procurement. Almost all government-owned discoms are saddled with long-term Power Purchase Agreements which now account for as high as 80% of the consumer bill. That leaves very little room for these discoms to manoeuver, severely restricting the scope for effecting significant reductions in power purchase cost and therefore, tariff. On the one hand, better quality of supply, another expectation from retail competition is also a chimera, since the competing supplier using the incumbent's network under open access is entirely reliant on the efficiency of the existing system.

There are far more reasons why retail competition cannot bring price or quality benefits to the Indian consumer. India's tariff design which incorporates steep levels of cross-subsidy has within it, the seeds of destruction of competition. In the absence of a cross-subsidy surcharge to compensate the incumbent for loss of customers cherry-picked by the new entrant, incumbent licensee is doomed to financial ruin. And when the regulator imposes a cross-subsidy surcharge as it eventually did (in addition to wheeling, banking and standby charges payable by the migrating consumer), it so erodes any price advantage to the new entrant, making migration unattractive to an existing consumer. With state regulators fixing cross-subsidy charges and other levies to protect the incumbent discom, competition through open access will succeed only where the new entrant has significantly lower-priced own power generation which is unlikely.

Besides, cross-subsidy surcharge (in open access cases) is not only complex to calculate accurately, but also difficult to enforce and require constant monitoring and vigilance. This is amply demonstrated by the number of tariff-related disputes brought to the regulator for adjudication.

(ii) **Licensees: Opportunistic and Contentious**

It is inevitable and even understandable that a provision like parallel licensing will invite opportunistic behaviour from the market participants. Predatory pricing practices to poach each other's consumers and garner market share, with a view to raising prices later should have been anticipated especially when there was a regulatory vacuum with respect to cross-subsidies for 21 months. Prayas report found that post-competitive entry, tariffs, after the initial reduction, went up, rendering illusory, any anticipated price benefit from competition (Prayas 2017) All we have is a see-sawing of consumers migrating between the competing licensees even as the regulator looks on passively without suo motu intervening to put in place the rules of migration ab initio. But even after introducing the protocols, there is little clarity as to whether a new connection should be deemed a parallel licence or open access, the latter attracting cross-subsidy surcharge.

Regulatory failure is writ large in the Mumbai retail competition experiment. All three discoms—TPC, RInfra and BEST—repeatedly petitioned the regulator, APTEL and eventually the Supreme Court itself, resulting in infructuous and protracted litigation which has still not reached closure after 18 years.

(iii) **Consumers—Demanding and Litigious**

Once the provision for consumer choice found its way into the statute, consumers readily jumped into the fray to avail the choice regardless of how it impacted the overall viability of the incumbent licensee. Some big consumers and consumer organizations moved the regulator and the higher judiciary to enforce their rights unmindful of the complexities involved in the process. This leads us to conclude that once choice is given, it will be seized. And once a statute confers choice, it cannot be denied to a demanding consumer. Therefore, ex-ante rules and framework should precede consumer choice.

(iv) **Government—The Silent Spectator**

The Government of Maharashtra, aware of emerging contestation in retail supply since 1998, did not act proactively to avoid duplication of networks in a crowded city like Mumbai. It could have done so under its powers to give policy directives to the regulator or it could have even modified the conditions of licence issued by it to the discoms. Instead, it chose to pass on the onus to the regulator (Prayas 2017).

(v) **The Regulator—Trial and Error**

Being the sectoral regulator with statutory authority and the necessary expertise to exercise it, much was expected of the regulator in terms of laying down the road map for introducing consumer choice. The manner in which the Maharashtra regulator handled the issues suggests that it was overwhelmed by the challenges of introducing competition in an industry traditionally considered a monopoly. Outsourcing it to the two competitors, directing them to engage a consultant to come up with suggestions on how it can be done is emblematic of regulatory failure to engage with emerging challenges.

Section 14 confers on the regulator, the power to decide when and where parallel licensing will be allowed. The words 'The Appropriate Commission may' in proviso 6 vests the regulator with discretionary powers to allow or not allow a parallel licensee to operate in an already licensed area. Perhaps TPC had a prior licence which it claimed, allowed it to distribute power to retail and small consumers, but it was up to the regulator to ensure the ground rules were in place to guide the transition. In allowing TPC to set up its own network in areas already serviced by RInfra, the regulator ignored the costs involved in duplicating a monopoly infrastructure. The regulator also failed to foresee the obvious deleterious impact it will have on the incumbent licensee. This undue haste in deciding that Mumbai was ready for retail competition set off an unhealthy competition between the licensees to the detriment of all stakeholders.

Section 86 (2) of Electricity Act 2003 empowers the State Commission to advise the state government on, inter alia, 'efficiency and economy in activities of the electricity industry'. Under these provisions, MERC could have even referred the matter to the Government of Maharashtra, seeking clarity on the ambit of the original licences issued to TPC in 1907 or advise it be amended. Yet, in adopting a passive approach, MERC allowed the situation to spiral out of control, necessitating intervention by the highest court.

To summarize, of all the stakeholders responsible for the messy outcome, the regulator must take the most blame. First, the regulator hastened to open the market without any ground rules in place. Second, the regulator does not seem to have anticipated cherry picking by the new entrant and mass migration by existing cross-subsidizing consumers to its network to the detriment of the incumbent; nor could it foresee predatory pricing strategies adopted by the discoms. Third, the regulator's decision does not display appreciation or acknowledgement of the uneven impact of regulatory asset charge on consumers who stayed with the incumbent licensee either because they chose to or they had no alternative. Regulatory intervention came much later and that too, at the behest of the affected licensee. Fourth, the regulator does not seem to have reckoned with the burden the incumbent bears as the supplier of last resort, as one with an obligation to supply to everyone on demand while the new entrant could pick and choose, although this situation was righted later; fifth, the regulator did not, ab initio, lay down conditions under which the new entrant should build its own network to service the migrating consumers as against those in which it should use the network of the incumbent licensee after paying applicable wheeling and other charges, in other words, clear entry criteria for market participants and a roadmap for parallel licensees. The changeover and switchover protocols which outlined some ground rules came too late to prevent the avoidable legal wrangles. And then, amidst allegations of cherry picking of consumers by TPC and in an attempt to balance the consumer mix, MERC imposed restrictions on migration, limiting it to small consumers and modified tariffs of both the licensees such that changeover became uneconomical for large consumers but lucrative for small consumers (Prayas 2017). Seventh, MERC failed to prevent escalating power purchase costs of the licensees, especially, RInfra which demurred signing long-term contracts and increasingly relied on short-term power purchase which turned out to be even costlier. Paradoxically, parallel licensing in Mumbai seems to have gone hand-in-hand with increasing costs for both the competing licensees and resulted in higher tariffs, something the regulator should have looked into, but did not. According to Prayas, none of the three discoms—BEST, RInfra and TPC—signed Power Purchase Agreements through the competitive bidding route, but settled for cost-plus purchases, often from own affiliates. (ibid) Scrutiny of such purchases is well within the remit of the regulator, but it did not exercise it in a timely manner. Finally, the regulator signally failed to address transmission constraints which further pushed up the cost of power as Prayas report has unequivocally established (ibid).

In this context, it is necessary to point out that MERC's two protocols—changeover and switchover—have given rise to further litigation which still have not reached closure.

MERC is signally responsible for the absurdity of allowing competition in a sector where both licensees were regulated and their tariffs fixed on a cost-plus basis. Cost-plus pricing is incompatible with a competitive market. Regulatory failure of serious proportions has come to define the Mumbai parallel licensing experiment.

(vi) **APTEL— Unclear and Inconsistent**

APTEL also had its share of inconsistencies in this unfolding saga of retail competition. In 2006, it interpreted the provisions of the various laws to concur with MERC that retail competition could be allowed, but at the same time, struck down TPC's claim that its original licence allowed it to engage in parallel distribution in Mumbai. It is interesting to note the completely divergent perspectives of MERC and APTEL on the interpretation of the same provisions.

Again APTEL differed with MERC on its interpretation of the rebates and discounts offered by RInfra. APTEL maintained that there was no prohibition on the licensee in the earlier laws of 1910 and 1948 or in EA 2003 from offering discount on the tariffs determined by the regulator. But bafflingly, it went on to add, on the basis of convoluted argument, that such rebates could be recouped by RInfra from the same consumers through tariff increases in future. This part of APTEL's order is egregiously violative of consumer interest since it paves the way for reckless predatory pricing practices by competing licensees within the same area. Such an observation is indeed corroborated by Prayas's report. Meticulously analysing relevant data, Prayas established that TPC raised its tariff in the next few years in a classic case of how predatory pricing eventually leads to higher consumer tariffs, triggering reverse migration to RInfra. This leads us to conclude that APTEL lost sight of long-term consumer interest when deciding on this dispute.

Yet, in 2012, when tangles created by operationalization of July 2008 Supreme Court order allowing consumer choice came up for adjudication, it was APTEL which took a considered stand and laid down clear guidelines differentiating the conditions under which new networks should be built as against those which could be under open access only. It also took the initiative to set up an institutional mechanism and put in place the modalities of assessing which route should be adopted, and if it involves duplication of infrastructure, the least cost option should be exercised. It is to APTEL's credit that it addressed tricky issues that even the Supreme Court failed to grasp and brought a modicum of clarity to operationalize consumer choice with minimum cost to the stakeholders. It is also to APTEL's credit that its 2014 order has not been challenged in the Supreme Court.

(vii) **The Supreme Court—Conflating Letter with the Spirit**

The Supreme Court engaged with all the technical and substantive questions raised by the appellants, but nevertheless, came to an erroneous conclusion in endorsing parallel licensing. The apex court conflated the provisions in EA 2003 authorizing open access with the provisions enabling parallel licensing which requires duplication of infrastructure. In that respect, the Supreme Court also failed to engage with the crucial question of whether it is prudent and necessary to duplicate electricity distribution network in any service area, and if so, under what conditions and circumstances.

The Supreme Court also chose to go by the letter, rather than the spirit of the law when it ruled that BEST, being a municipal utility, was exempt from providing open access through its network. It further ruled that consumers, however, are entitled to choice and hence could avail it only through duplication of network within BEST's licence area. This judgement does not consider the impact on the costs incurred in duplication of networks. Being the court of final appeal, the Supreme Court could have laid to rest all ambiguities in the laws through interpretation dictated by prudence and common sense. It lost this valuable opportunity, prompting an analyst to question whether the Supreme Court is indeed the appropriate forum for making such decisions (Daljit 2016). The slack left by the Supreme Court was eventually picked up by APTEL which laid to rest many ambiguities in the legal interpretation.

6.8 Analysis

Retail contestation in electric supply is a fraught battle world-over. In Mumbai, it has turned out to be a spectacularly costly experiment benefiting few stakeholders. The law is silent on the circumstances under which retail competition should be introduced in electric supply, entrusting it to the discretion of the sectoral regulator. The regulator in this instance put the cart before the horse, greenlighting retail competition without factoring in its obvious deleterious impact on the incumbent licensee. An alternative was available to the new entrant—'open access' to the incumbent's network. The regulator could well have taken a leaf out of the US Supreme Court's stand on a similar issue wherein the latter ruled that the feasibility of providing a common facility to the competitor should be exhausted before allowing duplication of that facility.

In the event, the premature and ill-considered measures to introduce competition in Mumbai's retail electric supply resulted in a messy situation in which unfair and restrictive trade practices were rampant and charges and countercharges were traded by a spectrum of stakeholders. The consequent legal wrangles could have been avoided had the regulator shown greater foresight, drawing up and declaring the rules of the game ahead of introducing competition. APTEL, a specialized sectoral appellate body too initially displayed an indifferent attitude towards consumer interest when it signalled, through its order, that predatory pricing was acceptable. Yet, in a later appeal, it was APTEL which acknowledged the wasteful nature of duplication of network and laid down detailed guidelines on how to differentiate when contestation should happen through open access and under what circumstances, a parallel network could be built. The Supreme Court, for its part, displayed its lack of familiarity with the sector and the peculiar ground realities in India when it operationalized parallel licensing. This leaves the question whether the sectoral regulator is better-suited to decide on complex issues that require in-depth and specialized knowledge.

Above all, operationalizing parallel licensing in Mumbai has led to an absurd and perverse outcome wherein competition, such as it is, emerged between players required to charge regulated, cost-plus tariffs, something that should baffle mainstream economic theories.

6.9 Chronology of Cases Relating to Mumbai's Retail Distribution

28.5.2020—APTEL order upholding consumers' right to choice of supplier; orders roll out of parallel network; Mumbai Network Distribution Assessment Committee to be set up

12.6.2017—MERC order on Changeover Protocol

28.11.2014—APTEL order on Migration Protocol

22.8.2012—MERC order restricting migration to small consumers; TPC appeals to APTEL against the restriction

11.1.2012—Competition Commission of India order on abuse of monopoly power by BEST

22.8.2011—MERC order on viability on account of migration in RInfra versus TPC

8.7.2008—Supreme Court order permitting retail competition

22.5.2006—APTEL order on Reliance Energy Ltd versus TPC

20.2. 2003—MERC order on Parallel licensing in TPC versus BSES.

References

Daljit Singh. 2016. Competition and Choice in Electricity Distribution in India, CPR Working Paper.
DEO. 2021. Delicensing Distribution, Lessons from Mumbai in Financial Express, Feb 10, 2021.
MERC. 2002. Order dated Nov 7, 2002 in Case 13 of 2002 filed by BSES for review of TPC tariff.
MERC. 2002. Order dated Oct 14, 2002 in Case 14 filed by BSES on violation of law by TPC.
MERC. 2012. Order dated Aug 22, 2012 in Case No.151 of 2011.
MRTPC. 2022. Order dated 28th Aug 2002.
Prayas Energy Group. 2017. In the Name of Competition.

Chapter 7
Regulatory Legitimacy and Turf Wars

Regulatory legitimacy is as much a function of the expertise of the regulators themselves, as the autonomy and independence accorded to independent regulators by a democratically elected government. When government takes away the authority conferred by the wisdom of parliamentarians, regulatory institutions are reduced to sinecures. Turf wars between the government and independent regulators are inherent in a democracy where basic utilities like electricity, cooking fuel and water are deemed merit goods with a political character.

The role played by governments is crucial to the success or failure of independent regulation. It is not inconceivable that governments, especially democratic governments, would be reluctant to cede their control over industries that supply basic utilities to apolitical and autonomous regulators. Consequently, governments have fashioned a toolkit with multiple tools to stymie the regulator and retain control of the sector. Despite the apparent autonomy and independence of the regulators, there are subtle and explicit ways in which the regulators can be reined in or even checkmated.

Among explicit controls wielded by governments is the power to issue binding policy directives to regulators, one that is written into the framework law itself. Governments also have the power to selectively withhold notification, (without assigning reasons), sections of the statute that brings regulatory commissions into existence in the first place. And the Indian government has wielded both these tools without compunction when it chose to. Governments can also take an adversarial position against regulators in courts when the latter are hauled up in disputes before the judiciary. Government law officers can and often do submit affidavits that contradict the spirit of the law. Occasionally, sections of the law are left vague enough, susceptible to multiple interpretations. Among the not-so-apparent tools are the financial clamps that prevent the regulator from hiring capable and competent human resources from the market to discharge their regulatory duties.

Yet, on the other side of the coin are 'rogue regulators' a term used to denote regulators, who by virtue of the authority entrusted to them by the statute, become

S. Mahalingam and K. Mahalingam, *Independent Energy Regulation in a Developing Economy*, https://doi.org/10.1007/978-981-99-4398-2_7

a power centre unto themselves, disregarding the interest of the sector and its stakeholders.[1] This is problematic especially when regulators, as in India, come from bureaucratic backgrounds and as such, are well-versed in ways not only to fight the turf war successfully, but also dodge built-in accountability mechanisms and emerge as parallel power centres to the detriment of the sector. A 'rogue regulator' could side with the consumers to the detriment of the entity, or champion the entity at the cost of consumers, defy or hoodwink accountability mechanisms and wreak havoc in the sector.

Against this backdrop, we examine the contours of two crucial disputes that throw light on both these aspects—In Part I, we discuss two diametrically opposite trends (i) government taking an adversarial position to stymie the regulator and (ii) the regulator attempting to become an unaccountable power centre. In Part II, we analyse a judgement of Delhi High Court to illustrate the turf war between the regulator and the government and the consequences thereof for the sector. In Part III, through an analysis of the same judgement, we examine how the Chairman of a regulatory institution managed to render ineffective, a collegiate multimember regulatory *body to concentrate power in his own hands. Judicial intervention was required to restore the collegiate character of the regulatory body. In Part IV, we discuss a policy directive issued by Government of India to an energy regulator marginalising and diminishing the legitimacy of the latter.*

7.1 Indraprastha Gas Versus Petroleum Regulator

Petroleum and Natural Gas Regulatory Board (PNGRB) Act was passed by the Indian parliament in July 2006. According to the preamble, the law is to regulate "the refining, processing, storage, transportation, distribution, marketing and sale of petroleum, petroleum products and natural gas excluding production of crude oil and natural gas so as to protect the interests of consumers and entities engaged in specified activities relating to petroleum, petroleum products and natural gas and to ensure uninterrupted and adequate supply of petroleum, petroleum products and natural gas in all parts of the country and to promote competitive markets and for matters connected therewith or incidental thereto".

The Board was established in July 2007. While the preamble is wide in scope, the remit of the regulator focused on natural gas infrastructure. Downstream oil sectors such as refining and marketing of petroleum products had been liberalized in 2002. Three major private investors such as Reliance, Essar and Shell had already set up refining and/or marketing facilities. With three existing state-owned refining and marketing majors such as Indian Oil Corporation, BPCL and HPCL, there was already a competitive market for the entire gamut of petroleum products.

---

[1] Rogue regulator is a term coined to denote regulators who become unaccountable and turn into a power centre unto themselves. It was used in 2008 in the context of FDA by Shorter Edward, 2008. Subsequently, the term was applied to power regulators by Tenenbaum in 1995.

Regulatory theory informs us that regulation is redundant where there are competitive markets since markets are expected to deliver the most efficient outcomes and regulation is only second best. Besides, petroleum refining and marketing are carried out largely without the aid of network infrastructure, since most petroleum products move in tankers through roads/railway wagons.[2] With little need of network infrastructure for transporting liquid fuels and the presence of competing players, petroleum product industry is already governed by competitive market forces, obviating the need for heavy-handed regulation. Therefore, PNGRB had limited role in liquid petroleum sector.[3]

However, natural gas is a different story. It requires a network of pipelines for transportation and distribution. As we have already seen, pipeline networks are natural monopolies warranting regulatory intervention. PNGRB's primary mandate, therefore, is to license new gas transport pipelines where there are none, so that over time, a nationwide gas grid develops to cater to consumers in all parts of the country. The regulator is also empowered to license city gas distribution networks since gas finds many uses in urban areas and the residential sector—as cooking fuel for domestic consumers, as relatively cleaner automotive fuel in the form of CNG, as processing fuel for industrial uses and as space heating and cooling for commercial consumers, etc. Of all its uses, Government of India has privileged its application in cooking, envisaging it would reach most residents in the country through a network of trunk pipelines and distribution networks that would be built over the next few years.[4]

When PNGRB came into existence in 2007, there were already a few existing trunk pipelines to carry gas from producing gas fields in Gujarat, Mumbai and Assam to industrial consumers in the north and west of the country. Natural gas pipelines also connected the two LNG terminals on the Western shores from where it was being taken to processing plants and consumer outlets. A few cities like Delhi, Mumbai, Ahmedabad, Baroda, Surat, etc. already had functioning city gas distribution networks, either owing to their proximity to gas fields or through deliberate policy.

In the event, the regulator's remit, mandate and powers focused on licensing new gas pipelines and city gas distribution networks (CGD), to extend the availability of this fuel to potential consumers in the rest of the country. The regulator is also mandated to facilitate non-discriminatory open access to existing trunk pipelines and city gas distribution networks and determine tariffs for such access and use. In fact, fixing tariffs of common carrier networks is a salient regulatory function assigned to PNGRB, and it is enshrined in Sections 20, 21 and 22. Specifically, Section 20 vests in the regulator, the power to declare a pipeline or local gas distribution network as a common carrier or contract carrier signalling open access.

---

[2] There are pipelines to carry petroleum products from the refinery to the marketing centres, and these are subject to regulation.

[3] Crude oil and crude oil pipelines were expressly kept outside the purview of PNGRB.

[4] The priority given to cooking gas is evident in the allocation of domestic gas expressly for supply to kitchens. Initially at 5 mmscmd, now all cooking demand has to be met from domestic gas production first by prorating reductions in supply to other consumers.

However, considering the substantial upfront capital required to set up piped gas distribution network in the cities, and the fact that they are natural monopolies in their respective licence areas, the regulator has the power to allow the pipeline or network builders exclusive marketing rights in their respective licence areas. (Anyone building the infrastructure would have exclusive rights to sell gas also in its jurisdiction) for a fixed period. Allowing exclusive marketing rights for a specified period of time would help the builder of the infrastructure recoup its capital investments, after which, the networks would be thrown open to competing suppliers for non-discriminatory access at a regulated tariff. The act envisages that the regulator would determine tariffs for all CGD/LDC networks to be charged to third-party suppliers after the expiry of exclusivity period.

In 2008, the Regulatory Board in the exercise of powers under Section 22 of PNGRB Act passed the Petroleum and Natural Gas Regulatory Board (Determination of Network Tariff for City or Local Natural Gas Distribution Networks and Compression Charge for CNG) Regulations, 2008, which would govern the procedure for determining the tariff of CGD networks for open acccess. These regulations, having been submitted to parliament and accepted, then acquired the force of law.

Indraprastha Gas Limited (IGL) had commenced city gas distribution in Delhi in 1999, having been hived off from GAIL, a trunk pipeline company, long before PNGRB came into existence in 2007. IGL was an integrated operation, sourcing gas from GAIL at the city gate station and distributing it to a mix of consumers in the city of Delhi. The gas consumers in Delhi received a consolidated bill for the cost of gas as well as the cost of distributing it within Delhi.

Since IGL had already been in existence for several years, it was time to end its marketing exclusivity and throw open its networks to competitors to ship their gas to consumers through IGL's pipelines to herald the benefits of competition. This meant that henceforth, not only could competing suppliers ship their gas to end-consumers through IGL's pipeline network in Delhi, but consumer bills would contain itemised break-up of price of gas, transportation charges, taxes etc. In 2012, PNGRB called for relevant data from IGL in order to fix network tariff and CNG compression charges. PNGRB then determined the Network Tariff and Compression Charges for CNG in respect of Delhi city gas distribution (CGD) network at Rs.38.58 per MMBTU as against IGL's calculation of Rs.104.50 and 2.75 per KG of CNG as against IGL's claim of Rs.6.66 per kg, to be effective retrospectively from 1 April 2008. The regulator notified these rates under powers conferred on it by Section 22 of the Petroleum and Natural Gas Regulatory Board Act, 2006 (PNGRB Act) and in accordance with its own Network Regulations notified in 2008.

The regulator also directed IGL to henceforth recover the said Network Tariff and Compression Charges for CNG separately through an invoice, without any premium or discount on a non-discriminatory basis. Relative to the network tariff fixed by PNGRB, what IGL had hitherto been charging (according to its own submission to the Board) was much higher. Therefore, PNGRB ordered IGL to appropriately reduce the selling price of CNG from the date of issuance of its order (1.4.2008) and asked the company to work out the modalities and time frame for refund of the differential in Network Tariff and the Compression Charges for CNG (in relation to what the regulator just fixed) to its consumers, with effect from the date of its order.

PNGRB would have to formally declare IGL's Delhi network as a common carrier before throwing it open to third-party suppliers. The network tariff and CNG compression charge notified in its order dated 1.4.2008 would take effect from the date of such declaration. Until then, PNGRB ordered IGL to provide the break-up of charges in its consumer bills—namely cost of gas at city gate, cost of distribution through its network and any other charges including marketing margin, etc.

Before the regulator could declare its networks as common carrier mandating open access, IGL promptly moved Delhi High Court, filing a writ petition under Article 226 of the Constitution of India, challenging PNGRB's powers to fix network tariff. In its petition before the High Court, IGL contended that it had been set up in 1998 under the directions of the Supreme Court to introduce a cleaner automotive fuel for public transport in Delhi and its environs under a central government licence to do so and as such, PNGRB did not have the power to fix network tariff nor throw open its networks to rivals in the name of open access.

IGL further claimed that PNGRB's order was tantamount to fixing the retail price of natural gas, interfering with its marketing freedom. The crux of the argument of the senior counsel for IGL was that Section 22 of PNGRB Act 2006 empowers the Board to only fix the transportation tariff and that the act nowhere empowers/ authorizes the Board to fix the price to be charged by a marketeer of gas from its consumers.

PNGRB, represented by the Additional Solicitor General, argued that the Board's order of 9.4.2012 did not fix the retail price of gas, but only the network tariff for distribution of the gas through the distribution network of IGL. And that its regulation titled Code of Practice for Quality of Service regulations sought break-up of charges in the bill in order to provide transparency in billing to benefit the consumers who have a right to know what they are paying for, especially when they are captive consumers of a monopoly transporter-cum-supplier. The regulator insisted that this was required to protect the interest of consumers and to foster fair trade and competition.

In its order dated 1.6.2012, the Delhi High Court held that PNGRB Act 2006 did not have any provision empowering the Board to fix the price or the Network Tariff or Compression Charges for CNG, as long as it is not transportation rate, (which is mentioned in the act) The High Court did acknowledge the mandate of the regulator to foster fair trade and competition, but in the case of entities distributing gas through their own distribution networks, the High Court held that the regulator cannot fix tariffs, nor can it fix the maximum retail price at which gas would be sold to consumers.

Whereupon, through a Special Leave Petition, PNGRB appealed to the Supreme Court challenging the defensibility and tenability of the judgement of the Delhi High Court. IGL, in its response, reiterated its earlier stand and questioned the powers of PNGRB to fix network tariff and that too when supplying its consumers through its own network. It alleged that the regulator was infringing its fundamental right to freedom of trade. The apex court had subsequently sought a response from the GoI, and the latter in an affidavit filed in July 2013 stated that the regulator had no power to fix the maximum retail price at which the entities will market or sell notified petroleum and natural gas.[5]

The apex court ruled that the function of the Board is of regulating the inter se relationship of entities under the act and not to regulate/control the relationship between the entities and the consumers. Similarly, Clause (f) of Section 11 of PNGRB Act 2006, while prescribing function of monitoring prices limits the same to taking corrective measures to prevent restrictive trade practices by the entities. Thus, only if the Board finds that the marketeers of gas in a particular area have formed a cartel or are indulging in any other restrictive trade practices, is the Board empowered to monitor prices. Such is not the case in the present instance. The apex court also held that price fixation/regulation/control is essentially a clog on the freedom of trade and commerce conferred the status of a fundamental right. Any restrictions on such a fundamental right can only be by legislative mandate which PNGRB did not enjoy.

The Supreme Court declared "We thus conclude that the PNGRB Act does not confer any power on the Board to fix/regulate price of gas as has been done vide the impugned order dated 9 April 2012", giving finality to the dispute over the remit of the regulator.

7.2 Section Analysis

The orders of both the Delhi High Court and the Supreme Court display lack of appreciation of the role of regulation in monopoly network industries in a liberalized economy. In a post-reform context, with government having withdrawn its oversight of these industries, such monopolies tend to price their monopoly service (in this case, transportation and distribution of gas) above marginal cost which is why they need to be regulated in the first place. PNGRB was set up to regulate the transportation and distribution tariff. In the instant case, the regulator was merely carrying out its mandate vested by PNGRB Act 2006.

While there are bound to be differences in the tariff calculations of the regulator and the entity, depending upon assumptions used for variables like capital expenditure, volumes, capacity utilization, etc. to outright deny the regulator's remit to perform its core functions is bound to have far-reaching implications for the sector; in fact, the judgement of the Supreme Court would lead to aggressive bidding for CGD in future where the winning bidder is free to charge whatever it chooses as long as it operates as a monopoly and maintain opacity in its billing.

Directing IGL to reduce its network tariff to the level determined by the Board and ordering refund of excess collected in this regard can only be in the interests of the consumers who are at the mercy of the monopoly supplier resorting to opaque pricing. Although IGL was allocated domestic gas whose price is controlled through the Administered Pricing Mechanism (APM) to be supplied to kitchens, the

[5] https://www.indiatoday.in/business/india/story/pngrb-indraprastha-gas-ltd-delhi-high-court-A-S-chandhiok-103090-2012-05-22 accessed on March 23, 2023.

domestic consumer had no way of knowing whether she was being charged only APM price for the gas she was using, thanks to consolidated billing. Separating gas price from its network tariff or distribution cost would bring in a level of transparency in billing, to the benefit of the consumers. But neither the Delhi High Court nor the Supreme Court acknowledged or appreciated this crucial role of the regulator. They resorted to semantics and the letter of the law to deny the benefit conferred by the spirit of the law.

Besides, the Board's orders were passed after following due process such as adopting the relevant regulations and having them approved by parliament in order to give them statutory force. In fact, IGL had submitted the data demanded by PNGRB, yet went on to challenge PNGRB's authority to fix tariffs, something that invited the argument of 'estoppel' by the Board's counsel.

As discussed in the chapter on theories of regulation, the raison d'etre of any regulation—by government or independent regulatory commissions—is to prevent network monopolies from charging prices above marginal cost, in other words, abusing its monopoly position to game the market. When fixing network tariff, the regulator was attempting to prevent precisely such gaming to serve the interests of consumers as well as improve overall allocative efficiency.

Three arguments have been advanced by the courts when coming to their conclusions. First, the regulator has no power to fix network tariffs of CGD entities except when it is 'transportation rate'. Second, as long as the CGD entity is supplying its consumers through its own network, it is not required to declare its network tariff. Third, to the extent the price charged to the consumer includes network tariff, PNGRB is indeed controlling retail price of gas which the regulator is not authorized to do so, under the provisions of PNGRB Act.

All three arguments are flawed. The term network tariff is essentially the same as transportation tariff. Perhaps, the Board should have used the term 'transportation tariff' which it did when it reissued the same regulation in 2020 (Petroleum and Natural Gas Regulatory Board (Determination of Transportation Rate for CGD and Transportation Rate for CNG) Regulations, 2020) substituting the term 'network' with 'transportation rate'. While not strictly adhering to the wording in the act, PNGRB was in fact, doing what it has been mandated to do, in the spirit of the law. After all, regulating the tariff of monopolies is the raison d'etre of regulation. IGL, in the present instance, was a monopoly charging prices above marginal cost according to regulator's calculations. The apex court could have ordered PNGRB to amend its regulation to use the correct phraseology and reissue its orders under the amended regulation. But it failed to do so. IGL could have appealed to the regulator for a review of the methodology and/or the assumptions used by it in arriving at the numbers, but instead succeeded in getting the baby thrown out with the bath water. Two decades after the judgement, none of the CGD entities have been declared common carrier to allow third-party access. Competition in gas supply, enabling which is a core regulatory responsibility, has been the casualty.

As to the second argument that an entity supplying gas to its consumers through its own pipeline network is not required to declare network tariff separately, this also is a deeply flawed reasoning. Regulation of monopolies is primarily to prevent dead-weight losses which can be caused by a monopolist charging higher than marginal costs. Since IGL was getting APM gas allocated by the government and the price of APM gas is transparently fixed, the consumer could now see for herself the price of gas, the cost of distribution to her premises (network tariff or transportation rate), and government taxes/duties if any, if her bill showed the break-up of each item. In the instant case, IGL was overcharging its consumer; its transportation rate per unit of gas was almost three times it should have been, according to regulator's determination done after due scrutiny of IGL's own data. All that PNGRB was trying to do was to pare it down to what it had determined according to a transparent methodology.

Even more egregiously flawed is IGL's argument that PNGRB was trying to regulate retail price of gas by regulating a component of the retail price. Even if IGL was supplying its own consumers through its own network, consumer interest is best served by fixing the distribution tariff rather than leaving it to the monopolist to decide whatever it chooses to charge. In fact, all components of the price of gas supplied to domestic kitchens was and continues to be regulated—APM price for cooking gas, regulated tariff for transportation, government-determined taxes and duties etc. Yet, IGL was charging an opaque price way above the regulated rates and managed to hide under 'own consumers through own networks' argument to escape regulatory scrutiny.

It is the conclusion of this study that both Delhi High Court and the Supreme Court of India have hit at the core of regulatory legitimacy through their perverse orders.

7.3 Chronology in PNGRB's Turf Dispute

1.7.2015—Supreme Court order on SLP dismissing PNGRB's claim

1.6. 2012—High Court concurs with IGL and rules that PNGRB does not have powers to fix network tariff. PNGRB moves SLP in the Supreme Court

9.4.2012—PNGRB issues order notifying Code of Practice for CGD, mandating break-up of items in the consumer bill. IGL challenges the order in the Delhi High Court invoking the courts writ jurisdiction.

1.4.2008—PNGRB notifies network tariff of Indraprastha Gas Limited

7.4 'Rogue' Regulators

In 2009, two writ petitions came up before the Delhi High Court for adjudication. One challenged the authority of the regulator on the ground that the Government of India deliberately withheld notification of a crucial section of PNGRB Act 2006, virtually emasculating the regulator ab initio, and rendering its actions invalid. The other writ petition challenged an order of PNGRB claiming that it was given by the then Chairman of PNGRB who had concentrated crucial regulatory powers in himself, bypassing the collegiate character of the multimember Board envisaged by the act. Both cases were heard by a bench of Delhi High Court which delivered a common judgement. An analysis of the arguments advanced by counsels for the main stakeholders—IGL, a city gas distribution company, PNGRB, the regulator, the petitioner NGO claiming to represent the public interest and the law officers speaking for the Government of India reveal as much about the turf wars between stakeholders as it does about the accountability of regulators themselves.

In 2009, two years after PNGRB was set up, Voice of India, a non-government organization approached the Delhi High Court with a writ petition, challenging PNGRB's power to issue licences since Section 16 of PNGRB Act 2006 had not yet been notified by Government of India. In the petition, Voice of India had made as respondents, not only the Union of India, but also PNGRB and IGL. It was filed in public interest under Article 226 of the Constitution challenging the illegal and arbitrary manner in which the affairs of PNGRB were run by its Chairman and seeking quashing of public notice issued by the Board inviting bids from interested parties for development of city gas distribution in Ghaziabad. Counsel for VoI also submitted that in the absence of non-notification of Section 16 of the PNGRB Act, the Board constituted under the act had no power to grant authorization to entities which had applied to the Board for "Laying, Building, Operating or Expanding City or Local Natural Gas Distribution Networks". Section 16 read with Section 2(d)(B) of PNGRB Act made it abundantly clear that authorization could only be given by the Board under Section 16 of the Act. He submitted that without the exclusive and monopolistic power being conferred on the Board, it had no power to issue any authorization to any entity.

Around the same time, IGL, the company that operated CGD in the national capital region too had filed a writ petition with Delhi High Court, challenging the authority of the Board to call for tenders to build CGD networks in Ghaziabad, (a town geographically contiguous to Delhi) for which IGL claimed was included in its licences for the National Capital Region issued by Government of India. IGL also challenged the Board's advertisements calling for tenders to build a CGD in Ghaziabad on the ground that the CGD licensing regulations have no force in the absence of notification of Section 16. Further, IGL questioned the authority of the Chairman in as much as the decision relating to Ghaziabad should have been taken by the entire Board alleging that the Chairman had bypassed the Board.

Both these petitions were taken up together by the Delhi High Court which sought the views of the Union of India on the powers of the Board to license trunk pipelines or CGD when Section 16 had not been notified. Mr. A.S. Chandhiok, learned ASG appearing for Union of India submitted that

> In view of non-notification of Section 16, it is most respectfully submitted that the view of the Central Govt. is that the Board is not currently empowered to issue authorizations for laying, building, operating or expanding any pipeline as a common carrier or contract carrier or city or local natural gas distribution network, as clause (d) (B) of Section2 of the Act defines an authorized entity inter alia is an entity authorized by the board under Section 16

of the Act. However, there is no impediment from taking preparatory work in this regard and applications can be processed till the stage or issuance of the Letter of Intent.

Relying on the ASG's submission, the Delhi High Court then proceeded to quash the Board's tender inviting bids for Ghaziabad on the ground that Section 16 had not been notified. The court declared that PNGRB did not have authorization powers and hence could not issue licences.

Voice of India's main contention in its writ petition filed with the Delhi High Court was that PNGRB's licensing-related functions were invalid because Section 16 of PNGRB Act 2006 had not been notified. Voice of India also questioned the validity of the actions of the Chairman in as much as he had rendered the Board redundant by delegating licensing powers to himself.

The following extract is from the Delhi High Court judgement dated 21.1.2010. It quotes the counsel for Voice of India "the affairs of the Regulatory Board had been turned by the Chairman into a one man show with other members virtually having no powers in day-to-day affairs of the Board". Important issues that should have been decided by the Board were discussed, if at all, in informal meetings in which no minutes or records were maintained. According to him, dissent or objections raised by members during these informal meetings were never recorded. He stated that when any issue was taken up in a Board meeting, no voting procedure was adopted although the PNGRB Act expressly provided for voting on issues on which there was disagreement; thus, all matters that came up before the Board were claimed to be unanimous even when there was disagreement among Board members. According to the same counsel, even written dissent notes circulated by the members were not taken on record. Further, as per Section 11 of the PNGRB Act, all the authorizations had to be given by the Board consisting of all its members and could not have been delegated to the Chairman. He submitted that the Supreme Court "in a catena of cases has held that the core/essential functions cannot be delegated by the Legislature".

The counsel for Voice of India further stated that the Chairman of the Board had delegated the powers of hearing complaints to officers of the Board. He stated that complaints had been heard and decided upon by the Secretary of the Board. Mr. Bhushan, the counsel further submitted that the Secretary of the Board was incompetent to decide the applications/complaints filed by the entities to the Board as Section 24(1) of the PNGRB Act specifically provided that any complaint/dispute referred to the Board had to be adjudicated upon by a bench comprising member (legal) and one or more members of the Board.

Section 24(1) reads as under:

24. Board to settle disputes—(1) Save as otherwise provided for arbitration in the relevant agreements between entities or between an entity or any other person, as the case may be, if any dispute arises, in respect of matters referred to in subsection (2) among entities or between an entity and any other person, such dispute shall be decided by a bench consisting of the member (legal) and one or more members nominated by the Chairperson. Section 58 of the PNGRB Act strictly prohibited delegation of power conferred under Section 24(1) of the PNGRB Act to any other authority. He submitted that despite this, most of the applications/complaints had

been heard and decided upon by the Secretary of the Board who after passing an order placed the same before the Chairman of the Board for his approval. According to him, the PNGRB Act nowhere authorized Secretary or any other officer of the Board to adjudicate upon the applications/complaints filed by the entities before the Board. This power was only vested with the members of the Board, it being their core function, and as such could only have been exercised by them.

These were countered by the Counsel for the Board claiming that since PNGRB had drawn up its authorization regulations in accordance with its own Conduct of Business Regulations and had obtained parliament's approval, these regulations had the force of law, the non-notification of Section 16 notwithstanding.

After hearing the counsels for both parties, Delhi High Court made the following observations: "Having heard the parties, we are of the view that it is through the principle of judicial review that the rule of law and other constitutional principles have to be given practical effect. Supervisory jurisdiction of a High Court has to be used to ensure that regulatory bodies, whether public or private, while performing public functions comply with the law and achieve acceptable standards of administration".

"Upon reading of the entire PNGRB Act, we are of the view that the legislative intent behind the PNGRB Act is to provide for establishment of a multimember Regulatory Board to regulate the refining, processing, storage, transportation, distribution, marketing, and sale of petroleum, petroleum products and natural gas so as to protect the interests of the consumers and entities engaged in specified activities relating to the said products and to ensure their uninterrupted and adequate supply in all parts of the country and to promote competitive markets".

"We are of the opinion that Section 16 is the source of power as it gives statutory mandate to the Board to issue authorizations. Section 16 also confers monopoly on the Board to issue authorizations. Without notification of Section 16, Board does not have the power to issue authorizations, inasmuch as there would be no ban on other entities from laying, building, operating or expanding CGD networks.

"We are further of the view that Sections 17, 18 and 19 of the PNGRB Act are all procedural sections in aid of Section 16. In fact, Sections 17 to 19 lay down the procedure to be adopted by the Board for inviting applications from entities and selecting the best among them. These sections do not give the Board the power to grant authorization to an entity which has applied to it. This power is specifically provided under Section 16 of the Act, and in absence of non-notification of the same, the Board cannot issue LOIs to any of the entities selected by it. If the respondents submissions were to be accepted, it would lead to chaos and would destroy the very object of the PNGRB Act which is to ensure that entities authorized by the Board are alone allowed to carry on the business of Natural Gas Distribution".

On the issue of collegiate decision-making the court observed "On a perusal of the PNGRB Act, we find that it provides for collegiate decision-making by a multimember Board after following the principles of natural justice. While dealing with an application for grant of authorization, several aspects including, legal, infrastructural, commercial, technical, etc. have to be considered. Consequently, PNGRB Act mandates that the Board shall comprise persons of eminence in the fields of petroleum and natural gas industry, management, finance, law, administration or consumer affairs.

"Even Sections 6 and 8(3) of PNGRB Act relied upon by the Boards learned senior counsel gives Chairman only the power of general superintendence and a casting vote. The PNGRB Act neither provides nor contemplates that all the powers and jurisdiction of five members of the Board would be concentrated in the hands of the Chairman alone—as has been done in the present case. In our view, the Chairman of the Board could not have delegated onto himself the core and essential function of the Board thereby rendering rest of the Board members redundant in the decision-making process. The Minutes of the 12th Board Meeting held on 11 September 2008 also do not indicate any ground to warrant delegation of such an essential function relating to the grant of authorization in the hands of one person".

"We are further of the view that the Chairman who was admittedly himself a delegatee, could not have further subdelegated his power to someone else—as he did in the present case by delegating his powers to the Secretary and the member (infrastructure). Section 58 of the PNGRB Act only envisages delegation of powers by the Board to a member or official of the Board by general or special order in writing. There is no scope for a delegatee to subdelegate that power". Finally, the court ordered the 'multimember board' to give a hearing to IGL and decide collectively.

7.5 Section Analysis

Regulatory theory lists impediments to effective regulation—absence of 'credible commitment' by governments to the new institution of regulation, asymmetry of information between the regulator and the regulated; inadequacy of accountability mechanisms to make reforms work and absence of commitment on the part of the regulator itself. And crucially, there is no level playing field that puts all stakeholders at par. For instance, hapless consumers are pitted against the might and wile of the government reluctant to cede turf, the machinations of the regulated entities that exploit the information asymmetry to benefit themselves at the cost of consumers and the autocratic tendencies of the regulators themselves to the detriment of public interest. Effective regulation is a delicate balance between an overbearing government and overly independent 'rogue' regulators. Such balancing is possible only if the accountability mechanisms such as collegiate regulation, transparent processes, parliamentary scrutiny and judicial appeals function as they should. That PNGRB members allowed their crucial power of deciding on licences to be hijacked by the Chairman speaks to lack of regulatory integrity and commitment on the part of the Board members and Chairman. The tendency to regard regulatory jobs as a sinecure calls for constant vigilance and prompt intervention by other stakeholders to correct the imbalance. Especially consumers who ultimately bear the cost of the regulated services have to be trained and equipped to exercise their rights by using the channels of accountability available to them. And finally, the judiciary must be able to grasp the complexities and nuances of the sector to be able to dispense justice.

7.5.1 Expropriation by the Government

Section 42 of PNGRB Act states

(1) The central government may, from time to time, by writing issue to the Board such directions as it may think necessary in the interest of the sovereignty and integrity of India, the security of the state, friendly relations with foreign states or public order.
(2) Without prejudice to the foregoing provision, the central government may, if it finds necessary or expedient so to do in public interest or for maintaining or increasing supplies of petroleum, petroleum products or natural gas or all or any of them or for securing their equitable distribution and ensuring adequate availability, issue policy directives to the Board in writing and such policy directives shall be binding upon the Board: Provided that no such directive shall relate to any day-to-day affairs of the Board: Provided further that the Board shall, as far as practicable, be given an opportunity of expressing its views before any directive is issued under this subsection.
(3) The decision of the central government whether a question is one of policy or not, shall be final.

In 2016, a decade after the Indian parliament vested the power of licensing in the hands of an independent regulatory body, Government of India unilaterally decided that a major gas pipeline covering 1836 kms and the CGD in the six cities/geographical areas lining the route of this pipeline would be built by GAIL, a gas transportation and marketing company owned by it, without going through a competitive bidding process to choose the most optimal bid. The Jagdishpur-Haldia-Bokaro-Dhamra Natural Gas Pipeline will traverse five eastern states of India including the city of Varanasi, the then Indian prime minister's parliamentary constituency. When completed, it is expected to yield political dividends to the party in power then, apart from forming a significant proportion of the nationwide gas grid. The cabinet committee on economic affairs even decided to fund 40 per cent of the cost of JHBDPL from the exchequer to take care of the viability gap GAIL might face in completing the project.[6]

On 17.11.2020, GoI issued another policy directive to PNGRB under Section 42 of PNGRB Act 2006, awarding Indradhanush Gas Grid Limited (IGGL), a joint venture of five public sector oil and gas companies, the licence to build and operate a northeastern gas grid that would take gas pipelines to the states of Assam, Meghalaya, Mizoram, Manipur, Arunachal Pradesh, Tripura, Nagaland and Sikkim.

A third-policy directive, entrusting supervision of the newly established Indian Gas Exchange to PNGRB was issued on 1.7.2020 although PNGRB Act does not mention gas trading or envisage supervision of gas trading exchange by the Board.

---

[6] Since this region has never consumed gas, there is uncertainty over finding enough consumers to warrant the investment. Hence, viability gap funding would minimize demand risk.

The government went ahead with the award of licences without reference to the regulator which under PNGRB Act is the sole authority to issue such licences.[7] JHBDPL pipeline authorization was issued under the prime minister's Pradhan Mantri Urja Ganga scheme which corroborates our observation that governments would like to retain control of major projects with anticipated political gains even if the law entrusts their licensing to the independent regulator.

While the act requires the government to consult the regulator before issuing such policy directives, there is no evidence that PNGRB was consulted by Government of India. It is common knowledge that competitively bid projects cost less than projects awarded on nomination basis.[8] Yet, GoI not only chose the nomination route to award such a major pipeline project to GAIL, but as bonus, threw in as many as six cities/ geographical areas along the pipeline to be developed into CGD networks. GAIL has also been offered a viability gap funding of 40 per cent of the project cost by GoI. Had this project been offered through contestation, such generous offers like viability gap funding (VGF) and the right to develop six CGD along the route, might have attracted private investors with the benefit of lower quotations. In fact, VGF itself could have been a bid parameter and the project could have been awarded to the party bidding the lowest VGF. Now, having awarded the project on cost-plus basis, the regulator is entrusted with the duty to fix tariffs of these pipelines. How public interest is served in awarding such a major project on cost-plus basis to a government entity without any contestation is a question that remains unanswered.

More importantly, by awarding this project through a policy directive to the regulator, GoI once again demonstrated its reluctance to cede turf to the regulator in deciding on a project that is perceived to bring political gains to the government of the day. In the process, GoI has weakened the regulator and put a question mark over its very legitimacy and competence.

7.6 Chapter Analysis

It is a matter of concern that more than two years after the Regulatory Board was set up; Government of India had failed to notify the crucial section of the act which empowers the regulator to carry out the core function for which it was set up in the first place. It confirms our hypothesis that governments set up regulatory institutions to satisfy Fund-Bank prescriptions, but are not committed to genuine regulatory reform. From the submissions made by the Additional Solicitor General, a senior law officer of Government of India, it is unequivocally clear that the omission was intentional, to hobble the regulator and to retain control of the sector. Coming from the High Court, the order lefts PNGRB with little scope for redressal of this grievous negation of its core regulatory authority. The perverse stand of the Indian government

[7] https://www.business-standard.com/article/companies/phase-1-of-pm-S-urja-ganga-project-to-be-commissioned-by-jan-gail-chief-118110600645_1.html accessed on March 23, 2023.

[8] Some good news in the power sector" Prayas in EPW Vol 42, Issue 2, Jan 13, 2007.

which resulted in the court order is testament to the attitude of the government to independent regulation in India's natural gas industry.

In the other instance discussed above, Government of India did not hesitate to exercise its political prerogative over the sectoral regulator by nominating GAIL, a PSU, to build a major trunk pipeline system in eastern India, covering, inter alia, the Prime Minister's own constituency. That this and the Indhradhanush projects have been awarded on cost-plus basis right under the nose of the sectoral regulator expressly set up to license such projects shows political expediency trumps norms of good governance, to the detriment of public interest.

Good governance is a delicate equilibrium of checks and balances. While creating the new institution of apolitical independent regulators, measures were introduced to ensure a modicum of accountability at every level. All regulations drafted and adopted by the regulator have to be arrived at after a consultative process with stakeholders; then these regulations will have to be placed in the parliament which scrutinizes them for conformity with the framework law. While implementing the regulations on the ground, the regulator is once again required to do it through a consultative process with the stakeholders concerned. When deciding on disputes brought before the regulator, it has to follow due process, conduct quasi-judicial hearings and issue speaking orders, setting out its reasoning. The media also keeps a watch over regulatory proceedings, providing an informal level of scrutiny. Finally, regulatory orders are appealable in the specialized Appellate Tribunal as well as in the higher judiciary when they concern questions of law.

Yet, despite these multiple levels of accountability, the then Chairman of PNGRB was able to marginalize the collegiate character of the Board to concentrate power into his own hands. Thanks to the provision for appeal, this subversion of regulatory process was arrested and reversed. In the end, the accountability of the regulator was restored and the checks and balances were tested and vindicated but not through regulatory initiative, but through the intervention of a consumer organization, pointing to the need to build strong consumer capacity to keep both government and regulator in check.

Chapter 8
Summary, Findings, Conclusions and Policy Implications

This concluding chapter summarizes the findings from this research, the conclusions that emerge from the findings and the policy implications thereof and analyses them against the backdrop of mainstream regulatory theories developed in the context of advanced economies.

Three decades ago, in the aftermath of two successive oil crises, governments of the world affected a paradigm shift in the way their economies were being governed, introducing a slew of comprehensive macroeconomic reform measures that envisaged a migration from state to market in most spheres of the economy, including basic network utilities such as electricity, water supply, road networks, hitherto organized and supplied by the state directly or through its agencies. As part of this process, network utilities were unbundled and restructured into corporate entities and privatized where feasible; a new institution of independent regulation was established for most utility industries avowedly to improve efficiencies by introducing competition where feasible and mimicking the market where not. The independent regulator is indeed at the core of the new governance paradigm, empowered by statute, entrusted with appropriate remit and invested with adequate authority. Tasked to oversee the transition from state to market, the regulator would play the role of an impartial and objective umpire providing level playing field to new entrants into the market.

In this research, regulators have been singled out for scrutiny not only because they are a new governance institution, one that does not readily fit into the conventional Madisonian system of checks and balances, but also because their competence and performance are critical to the outcome of the reforms. Thirty years is a long enough period to undertake an assessment of the effectiveness of this institution in delivering the goals for which it was set up.

Regulators are unelected and as such are subject to a somewhat intangible and diffuse web of accountability mechanisms not all of which are equally effective.

Supplementary Information The online version contains supplementary material available at https://doi.org/10.1007/978-981-99-4398-2_8.

S. Mahalingam and K. Mahalingam, *Independent Energy Regulation in a Developing Economy*, https://doi.org/10.1007/978-981-99-4398-2_8

Regulators' accountability is derived from parliamentary scrutiny of the regulations they draft, the government's overarching policy framework within which they must operate, the transparent and participatory manner in which they conduct their day-to-day business and the appellate jurisdiction of the higher judiciary to which their adjudicatory orders are subject. Above all, regulatory legitimacy arises from the regulators' own expertise—real or perceived—on the matters they regulate. In fact, expert knowledge and insights which they bring to bear on the sector they regulate are at the core of regulatory legitimacy. In other words, regulation is equated with governance by experts. Therefore, this research also seeks to examine whether such expertise is reflected in their functioning.

In the Indian context, electricity regulators have the hapless task of regulating government-owned monopoly utilities without the majesty of government authority. Being creatures of government, they are viewed as being less legitimate than governments themselves (although a truly independent regulator is not an agent of the government and hence not in a principal-agent relationship). Perhaps, this explains their diminished credibility and effectiveness in regulating government-owned entities. That, however, goes for their regulatory role—of rule-making, licensing, tariff determination, monitoring of implementation of licence conditions, etc.

Indian energy regulators have an additional role—as quasi-judicial arbiters of disputes between stakeholders in the sector, the first port of call, so to speak, for feuding stakeholders. In this role, however, regulators step back from the overarching control of governments (through the latter's power to issue policy directives etc) to come into their own although here too, they are subject to appeals to the higher judiciary. Since they are not creatures of the judiciary, their adjudicatory role has a salience all its own, not least because of their (perceived) domain expertise.

The adjudicatory role of regulators is salient also because their quasi-judicial orders reveal regulators' understanding and appreciation of their own role in a larger liberalized economy. After all, regulators play a crucial role in the evolution of jurisprudence in a liberalized economy. Although, even here, ultimately, their understanding and wisdom on issues brought before them for adjudication are tested against the (perceived or real) superior wisdom of another 'expert' body, the sectoral appellate authority (APTEL), and in many cases, the higher judiciary as well. The latter being part of the Madisonian system of checks and balances does indeed have greater legitimacy and finality. Yet in a complex technology-driven modern economy, an expert regulator is expected, by virtue of its domain expertise and the wherewithal, to shape jurisprudence in furtherance of the reform agenda. This is the hypothesis that is tested by this research.

Regulatory theory informs us that economic regulation is warranted only where there is a monopoly. Competitive markets regulate themselves effectively. Yet we have a situation where the Indian electricity regulator is called upon to regulate a hybrid arrangement—a partially privatized and competitive segment (power generation) that feeds into a largely government-owned monopoly transmission and distribution segment. All of them must coordinate and function harmoniously to deliver electricity to the final consumer. Herein lies the challenge.

In India, as elsewhere in the developing world, there are many layers of complexity not foreseen in the Western context where many theories of independent regulation evolved. For instance:

(i) In the hybrid model, generation costs are not effectively regulated but are a pass-through while transmission and distribution tariffs are regulated.[1] This has led to a somewhat anomalous situation. Power purchase costs have fallen out of ex-ante regulatory scrutiny, but disputes bring them back to the regulatory review, ex-post.
(ii) Regulators are called upon to introduce competition in an industry where consumer prices include an element of cross-subsidy for an undifferentiated commodity like electricity. To illustrate, industrial and commercial consumers pay higher tariffs which cross-subsidize agricultural consumers who pay zero tariffs in some states and very low tariffs in others. This vitiates the scope for competitive markets which the regulator has been mandated to introduce.

Which is perhaps why, unlike most of their Western counterparts, Indian energy (both electricity and petroleum) regulators have been endowed with quasi-judicial powers to settle disputes between stakeholders in the industry.[2] Ever since they have been set up, India's energy regulators have been called upon to adjudicate a plethora of disputes which take up much time and resources, far greater than those required to discharge their regulatory functions.

This inquiry is focused on that singular area of regulatory performance, namely the quasi-judicial role of India's electricity regulators—not only to evaluate their dexterity in handling this complexity, but also to get a sense of their own understanding and appreciation of the role they are expected to play in the evolution of jurisprudence in liberalizing economy. Being disputes involving high-value transactions, naturally, most of them get appealed to higher judicial forums—APTEL—and eventually end up in the highest court of the land. This research tracked the evolution of jurisprudence through landmark disputes related to key issues salient in a liberalizing economy—sanctity of contracts, interpretation of contracts, retail competition, etc., through all adjudicatory levels, starting at the regulator right up to the Supreme Court of India. The objective is to understand the perspectives of expert regulators vis a vis those of general judiciary, in shaping jurisprudence in a liberalized economy.

From our analysis, India's energy regulators do not seem to reveal any consistent pattern across time or space when exercising their quasi-judicial function. In some instances, regulators have displayed a greater appreciation of their role in a market economy than the higher judiciary, and at other times, regulatory orders appear whimsical. But the main hypothesis of our enquiry, namely whether regulators being domain experts, have a better appreciation of their role than either bureaucrats or the judiciary, does not seem to have conclusive answers. Whether this is a reflection of the level of regulatory expertise or otherwise is a subject matter for future research.

[1] Competitive bidding for power projects is believed to obviate the need for regulatory intervention in the generation segment which is why power generation costs are a pass-through in the tariff.

[2] Individual consumer disputes are not heard by the regulator. Only class action suits are.

In the following section, we summarize our findings from each of the chapters dealing with landmark disputes adjudicated by energy regulators.

8.1 Sanctity of Contracts

On sanctity of contracts which dealt with three major landmark disputes, CERC the central electricity regulator rightly upheld the sanctity of contracts in a market economy. It unequivocally expressed its disapproval of reopening validly concluded contracts. Yet, at the same time, in a manner tantamount to bypassing the sanctity of contracts, the regulator found a way to compensate the power developers through compensatory tariffs, citing financial viability of the generating stations as an important consideration to enable them to continue to supply power to the consumers.

It is indeed a tricky balance between the viability of the generator and the welfare of the consumer (by keeping tariffs from rising). Yet, the regulator's order seems to tilt in favour of the developer in as much as it eventually ended up ordering compensatory tariffs, invoking the special powers conferred by the statute. In the process, it makes a mockery of the concept of 'sanctity of contracts' even as it pays lip service to it. The regulator's order dwells in detail upon the hardships faced by the power projects on account of developments claimed to be beyond the latter's control. The central regulator justifies award of compensatory tariffs on the plea that calling for bids to set up new projects (to compensate for the cessation of supply from the disputed ones) will be even costlier. Such an argument will almost always hold good by sheer dint of passage of time between bids. Future bids are always likely to throw up higher tariffs than past ones, unless there are disruptive technologies or new resources available to future bidders, which is not the assumption made in the order. Hence, it cannot be a valid argument for intervening in market outcomes. This contradiction is glossed over in the service of pragmatism. A truly competitive market for which the regulator is deemed a surrogate, does not accommodate business risks gone awry to deliver pragmatic outcomes.

This raises two important questions: (i) whether markets with their rules of free entry and exit are indeed the ideal institutions to supply an essential commodity like electricity because exit is not really an option; (ii) whether hybrid markets like the Indian electricity industry are amenable to same market rules applicable to fully competitive markets.

The regulator seems to have relied heavily upon the recommendations of an external 'expert' committee appointed by the stakeholders themselves to find a way out of the impasse. While acknowledging the slender resources available to the regulator to investigate the claims of the petitioners, here was a golden opportunity for the regulator to seize and apply its 'expertise' and establish its own credibility and legitimacy and yet, it relied on external inputs, that too from interested parties to come to a decision on such a core issue relevant to its very raison d'etre. It tempts us to conclude that perhaps the expertise of the regulators is more perceived than real.

The regulator also failed to question the developers as to why, when they had the option to factor in fuel risk through the 'escalation clause' in the bid document, they chose the fixed price option to supply power fixed for the next 25 years. It is well within the regulator's remit to point out that the power developers knowingly took a commercial risk and hence should be held down to fulfil their commitment. It was incumbent upon the regulator to point out that contract renegotiation or award of supplementary tariffs is prejudicial to the bidders who lost the bid ab initio, because they quoted higher (and perhaps more realistic) tariffs and hence lost the bid. Neither of these arguments finds mention in the regulatory order.

Two lessons emerge from this case study: first that regulators are in agreement with the view that markets know best and market-determined prices should not be tampered with. It explains their hesitancy to tamper with competitive outcomes especially when the law is not unambiguously on their side.[3] Yet, the acute information asymmetry between the private developer and the discom throws up perverse outcomes even in a competitive setting, bringing them back to regulatory scrutiny ex-poste in the form of disputes to be adjudicated upon. Even here, regulators seem to delegate their responsibility to external experts and committees. Our research suggests that independent regulators in India are not at ease, occupying the rightful space assigned to them (by virtue of their expertise) in the larger governance framework. Whether it owes to their self-perceptions of their own domain expertise or their pecking order in the Madisonian system of checks and balances could be subjects for future research.

Second and most crucially, it confirms our view that regulators are accountable only for their acts of commission, but not of omission. Both these findings have policy implications.

The Hinduja project in Andhra Pradesh reveals lack of credible commitments both on the part of the government and its entities. News reports point to issues relating to sovereign guarantees of Government of India as the cause of 18-year delay in commissioning the project, leading to cost overruns. In the absence of credible information on the actual causes of delay, this research refrains from commenting on this aspect. However, the flip-flop policies and prevarications of the state discoms display a lack of credible commitment on the part of the state and its agencies, exacerbated by delays and confusion arising out of the bifurcation of the state. The private developer, for its part, seems to have taken advantage of the cost-plus status of its project to present the discom with bills for cost overruns, ex-post, leading to a stalemate in which the Supreme Court intervened and castigated the regulator for allowing the discoms to stop drawing power from the project.

This is another case study which reveals the lax accountability of the regulator. As observed in the earlier case study, here too, the regulator absolved itself of responsibility to scrutinize project costs (a core regulatory responsibility under Section 62 of Electricity Act 2003) by an act of omission—omission to insist on regulatory

---

[3] The Supreme Court in its order dated 2 February 2022 held categorically that the general regulatory powers conferred under Section 79 of Electricity Act 2003 would override other prohibitions in the law.

scrutiny of project costs and determine tariffs. Regulatory act of omission is also evident in the fact that it failed to safeguard the interest of the project developer with sunk investments which cannot be moved elsewhere. By not ordering performance of the contract (PPA) as prayed for by the developer, the APERC created a classic 'hold-up' problem, inviting censure by APTEL as well as the Supreme Court. It leads us to conclude that the expertise of regulators is only as good as its application in regulatory performance.

The policy lesson that emerges from this case study points to the need to make the regulators accountable not only for their acts of commission, but also for their acts of omission—for not performing their mandated duties critical to the functioning of a liberalized market economy.

8.2 Implying Terms into a Contract

When Adani, a private power developer unilaterally terminated its PPA with a Gujarat discom and stopped supplying power to the State of Gujarat, the Gujarat regulator rejected Adani's claim to imply terms into the contract to accept the termination. Instead, GERC ordered resumption of power supply as per the PPA. In coming to this conclusion, the regulator relied not only upon the claims made by Adani in its petition, but marshalled evidence from other statutory documents like filings with SEBI and the Annual Report of the company. This is an instance wherein the regulator not only used its expert knowledge of the sector but also conducted due diligence based on which it came to the conclusion that the termination of contract was unjustified. APTEL also rejected the appeal by the developer albeit on a different ground.

But the Supreme Court chose to insert terms into the contract, citing business efficacy. In doing so, the Supreme Court not only contradicted its own verdict in another case where it had held that the 'Penta' principle must be satisfied before implying terms into a valid contract but opened up the ground for similar demands in future, vitiating the sanctity of contracts in a liberalized economy. Further, perhaps unwittingly, the Supreme Court had also ignored the liability of the party responsible for the default, namely GMDC, and foisted it on the hapless consumer by ordering compensatory tariffs. In a business involving several links in the value chain, the liability of each actor must be clearly defined in the contract itself and enforced, a point that the apex court failed to address.

Our case study on implying terms into the contract shows that the regulator demonstrated its expert insights in rejecting the demand for implying terms into a validly concluded contract whereas the Supreme Court, by setting aside the orders of the regulator and APTEL, not only overlooked the impact of tariff increases on the consumers, but also weakened the authority and legitimacy of the regulator.

On the issue of interpretation of the provisions of PPAs, the Punjab regulator also failed to fix liability on the defaulting party in a value chain consisting of back-to-back contracts. While rejecting all claims made by a private power developer—such as coal-washing charges, transportation charges, costs on account of loss of calorific value of coal, etc. the Punjab regulator could have pointed out that compensation for freight costs should be claimed by the project developer from the railways which delayed construction of a siding on account of which the developer had to ship coal by road. Instead, this was disallowed by the regulator. That the liability arising on account of delayed construction of railway siding should lie with the railways, not with electricity consumers is not an argument articulated by the regulator.

APTEL, on the other hand, allowed costs incurred by the developer (on account of delay by the railways) to be passed on to the tariff, transferring the burden on to the hapless consumers. The Supreme Court, for its part, granted many of the claims made by the developer over and above the terms of the contract, further burdening the consumers.

This leads us to the conclusion that in India, contracts are open to diverse interpretations. The lessons learnt from claims arising on account of expenditure not factored into long-term PPAs must be used as a learning curve to fine-tune future PPAs. Back-to-back liability needs to be specified ab initio in the contract, especially for large projects involving coordination among many stakeholders along the value chain. A liberalized market economy is expected to allocate penalties as efficiently as it allocates resources. Regulators need to be sensitized to this reality.

8.3 Regulator as Market Maker

Promoting competition in electric supply and mimicking the market until competition arrives is at the core of regulatory remit. Yet, in a traditional monopoly business, this is a task for the expert and needs to be sequenced right so as to avoid manipulation of the market by the contestants. In the Mumbai case study analysed in Chap. 6, the regulator appears to have been overwhelmed by the daunting task of introducing competition. Even when competition originated suo moto, the regulator, instead of seizing the opportunity to proactively lay down rules of the game for orderly roll-out of retail competition, chose only to react that too, only when presented with a dispute for adjudication. While the ambiguity in the licences issued to the competing discoms also contributed to this confusion, neither the government nor the regulator stepped in to unravel the tangle that resulted from unsupervised and unregulated development of competition in a traditional monopoly business. The resulting chaos was rife with allegations of market manipulation, predatory pricing and anticompetitive behaviour by the contestants. Yet, instead of dealing with the dispute, the regulator resiled from its own responsibility of laying down the rules of the game and instead, ordered the stakeholders set up a consultative mechanism to thrash out the modalities of contestation. Such an order does not inspire confidence in the expertise of the regulator.

While the law does allow for parallel licensing of retail suppliers involving duplication of networks, a sectoral regulator with expert knowledge of the problems peculiar to a developing country context such as the presence of cross-subsidies, is in an ideal position to decide how, when and where it should be introduced as well as its sequencing. Instead, in Mumbai, competition developed in a chaotic manner. Unable to get the regulator to lay down the ground rules, the parties approached APTEL which too washed its hands off initially. The Supreme Court, lacking the expert insights of a sectoral regulator, further complicated the situation, conflating open access with parallel licensing. Finally, it was left to APTEL to lay down detailed ground rules for migration of consumers from one supplier to another. Even now, despite the ground rules, it is an experiment in the making.

The Mumbai case study confirms our observation that regulation as a governance mechanism does not command legitimacy also because it does not display the expertise expected of a sectoral regulator. Retail competition fraught with complexity is a fit case for the domain expert, not the general judiciary. Yet, neither the regulator nor the apex court could grapple with the complexities, once again leading us to question the theory that refers to regulation as 'rule by experts'.

8.4 Turf Wars

Any new governance institution, pitted against the incumbents has to fight for its turf. Independent regulators, even though endowed with statutory authority, are no exception. In Chap. 7, we examined two case studies, one where this turf war between the government and the regulator played out to the detriment of the latter, striking a blow to regulatory authority and the other, where a rogue regulator was chastised by the Delhi High Court, weakening its legitimacy.

In the first case, we examined how IGL, the city gas distribution utility in the national capital defied the regulator's attempts to introduce competition in retail gas supply. IGL refused to submit to the regulator's orders to fix network tariffs nor allow open access to its monopoly network to competing suppliers. IGL challenged the regulator's authority to do so that too under the writ jurisdiction of the high court with its wide amplitude. The Delhi High Court not only entertained the petition under its writ jurisdiction but even held that the regulator did not have the power to fix network tariff, putting paid to the introduction of competition in gas distribution and striking at the root of regulatory raison d'etre in a liberalizing economy.

Unlike the other cases examined in this book, in this case study, the regulator was not adjudicating any dispute but itself became a respondent in a petition filed by one of its regulated entities. It did so to challenge a regulatory order on tariffs when it had the option to approach the regulator with a review petition. Invoking the writ jurisdiction of the High Court, designed to redress weightier constitutional matters to settle a commercial dispute should not have been encouraged in the first place. Yet, the High Court entertained the petition and ruled against the regulator.

That the regulator was dragged to court by a regulated entity should itself be a cause for concern since it displays contempt for a statutory institution set-up under an act of parliament. It is also a matter of concern that the Government of India which was instrumental in getting the statute enacted in parliament and in setting up a Regulatory

Board to oversee the transition of the sector from monopoly to a contested business, ds took an adversarial position to the regulator. GoI had maintained that the intent of the PNGRB Act 2006 did not contemplate empowering the regulator to fix network tariffs, striking at the root of parliamentary intent.

The High Court order struck at the very root of reforms in the petroleum industry and weakened the regulator's role, remit and authority. Its ruling was not in consonance with the ethos of a liberalized economy where the regulator has been expressly installed to promote competition in industries hitherto dominated by monopoly franchisees. Setting network tariff is a necessary precursor to introduction of competition in retail gas supply. Yet, based on specious arguments made by IGL's counsel conflating and equating tariff determination with price control, the High Court restrained the regulator from introducing competition in retail gas distribution. When the regulator appealed to the Supreme Court, it too upheld the High Court ruling displaying lack of appreciation of the regulator's role in a liberalizing economy.

The takeaway from this case study is that perhaps the courts should exercise a degree of restraint in matters involving domain expertise, by not admitting complaints that can be better dealt with by experts. The courts would also do well to be cognizant of turf wars ineluctable in a regulatory paradigm.

8.5 'Rogue' Regulator

Regulatory institutions in India have been designed as multimember bodies—at least three members in SERC, four in CERC, and five in PNGRB, including the respective Chairmen, each selected for his or her expertise in matters such as engineering, finance, law, economics, commerce, etc. The law envisages collegiate functioning of the commissions/Board under administrative superintendence of the Chairperson. Regulators, unelected and appointed for a tenure during which they cannot be easily removed, enjoy wide-ranging remit and powers. They are held accountable also by their collegiate functioning.

The second half of Chap. 7 relates to appropriation of Board powers by the Chairman of PNGRB through illegal delegation. A consumer organization approached the Delhi High Court with evidence on how this was done, following which the High Court struck down the concentration of powers into the hands of the Chairman of the Regulatory Board and ruled that all decisions are envisaged to be collegiate by law.

The same case study also points to how governments can stymie a statutory institution, by using its powers to selectively notify sections of the framework law under which the regulatory institution is set up. When asked by the Delhi High Court for the government's stand on this, the government counsel stated in court that in view of non-notification of the concerned provision, the Board could not perform one of its key functions, namely licensing of pipelines and CGD networks. In the context of PNGRB set-up expressly to expand the pipeline networks and CGD in the country, non-notifying licensing powers is tantamount to emasculating the regulator.

The provision was notified by the government subsequently, three years after PNGRB was set up. It is a classic case to show how turf wars are inherent in the regulatory paradigm. It also points to the possibility that there is no buy-in from the government which has been persuaded to open up basic utility industries to private capital under the oversight of an independent regulator.

In yet another case study reviewed in the same chapter, we find that even after notifying Section 16 of PNGRB Act 2006 which confers licensing powers on the regulator, the government used its powers to issue policy directives to wrest control of the same licensing functions for a major project covering both pipelines and CGD in eastern India.

We conclude, therefore, that independent regulation is a reluctant institution in developing economies, thrust on host governments by Fund-Bank prescriptions rather than emerging from a felt-need from the grassroots. It also points to the political nature of energy industries which governments are reluctant to hand over to apolitical regulators.

8.6 Conclusions and Policy Implications

The case studies in this research point to the fact that current conceptual frameworks of regulation devised in the context of the developed economies sit ill in developing country contexts. This research finds that regulating government-owned utilities is a tricky issue since the government is always higher in the pecking order of legitimacy and authority. While the regulatory model was envisaged in the anticipation of privatization of distribution utilities, ground realities have forestalled privatization in India leaving us with a perverse situation wherein the hapless utility is subject to two seemingly adversarial masters. The adversarial nature of the relationship stems from the short-term goals of politicians while regulators have no such compulsions and hence are expected to regulate for the long-term health of the sector. Yet, regulators installed to depoliticise decision-making in the sector are effectively thwarted by politicians, who, in their role as owners of the utility steer it in the opposite direction, also because the commodity they regulate has political economy characteristics. This leads us to conclude that the institutional context is crucial to designing a regulatory framework for developing countries.

Estache and Wren-Louis, developing on the original model of Jean Jacques Laffont, have attempted such a design, assuming four limitations in developing countries, namely (i) information asymmetry, (ii) limited regulatory capacity, (iii) limited commitment and (iv) limited accountability. Regulatory outcomes are affected by such limitations leading to, inter alia, adverse selection in auctions. "The contract may not go to the most efficient firm, but instead to the one that believes it has the highest chance of renegotiating (or the lowest chance of expropriation)" (Estache and Wren-Lewis 2009). All these limitations are present in India which is probably why there is demand for renegotiation of a large number of contracts.

As Demsetz would have us believe, information asymmetry can be circumvented through the auction process, but as we have seen in the Indian case studies, auction alone is insufficient (Demsetz 1976). Limited regulatory capacity is evident in many of the adjudicatory orders examined in this research. Commitment remains a problem throughout the developing world, not least because of resource constraints of governments, but as our case studies show, lack of credible commitment could also be from the project developers, not always on the part of the government.

Regulatory accountability is largely exercised through parliamentary scrutiny of delegated legislation, participatory regulatory hearings and speaking orders of regulators which explain the rationale for their decisions, but also through the appellate jurisdiction of APTEL, the Supreme Court and the writ jurisdiction of the higher judiciary. The judiciary, however, does not boast of domain expertise and sometimes gives judgements not in consonance with the ethos of a liberalizing economy.

From our case studies, the following recommendations emerge:

(i) Hybrid electricity markets (part-private and part-government-owned) are not amenable to effective regulation or competition. Rules of competitive markets characterized by free entry and exit do not apply to hybrid markets, especially one supplying a basic necessity like electricity or water. Hence, a new governance paradigm is needed. Unbundling need not be a prerequisite for competition. Vertical integration is perhaps a better way to organize electric power industry in developing countries. This is also borne out by the profitability of vertically integrated private utilities in India. Breaking up existing franchises into smaller units, each vertically integrated, is perhaps one option for the future. Regulators can signal competition between franchisees through appropriate incentives.

(ii) Power purchase costs which now account for 60–85% of the final tariff in most states need to be brought under regulatory purview through an amendment to Electricity Act 2003.

(iii) Expertise of regulators cannot be assumed. Regulatory capacity needs to be created and strengthened through dedicated training efforts. Regulatory studies should be integrated into the university curriculum.

(iv) While acknowledging that long-term contracts can never be complete, it is possible to build-in greater flexibility in PPAs to allow for unforeseen circumstances outside the control of either party. Perhaps, GoI guidelines could be revised to allow only for 'escalable' tariff contracts, and apart from the quoted initial tariff, the rates of escalation could be another biddable parameter. This leaves the regulator to scrutinize only the genuineness of claimed escalation and the correctness of the quantum of compensation sought. But once the escalation clause is agreed upon, no further renegotiation should be allowed during the life of the project except for the reasons included in the contract such as 'Force Majeure' and 'Change in Law'

(v) Some of the factors that have been agitated before the regulator and the courts praying for compensatory tariffs over and above that agreed in the PPA are not exactly unforeseeable and should have been clearly spelt out in the PPAs ab initio. Future contracts must be tweaked to include all the lessons learnt from the extensive litigation that has plagued the existing PPAs.

(vi) Perhaps, it is not a good idea for any state to rely predominantly upon a single large power project for its electricity needs wholly or substantially. While there are undeniable scale economies in large-scale power plants, discoms can avoid the 'hold-up' problems brought on by demands for renegotiation of contracts by diversifying their source of supply so as to escape being held hostage by projects 'too big to fail'. Perhaps, there is a case for limiting the size of exposure of a State to a single supplier.

(vii) Currently, the two-part tariff policy saddles the buyer with fixed cost for the life of the project. Perhaps, the time has come for moving away from this paradigm towards merchant plants and/or develop a wholesale power market. Appropriate incentives in the form of land allocation at discounted prices could attract investments in generation.

(viii) Perhaps, there is a case for involving the regulator in the competitive bid process either at the bid design stage or at the approval stage. CERC has formulated a benchmark for MoU-based projects. Similar benchmarks for competitively bid projects with due consideration for technology, fuel, location, etc. might be useful for the regulator to evaluate the feasibility of even competitively bid projects. This might need amendment to the law.

(ix) The institution of regulation should expressly include provisions for effective consumer participation in regulatory proceedings. The absence of consumers from the regulatory paradigm is egregious and damaging to consumer interest.

(x) Electric supply industry requires coordination by a range of actors such as coal miners, coal suppliers, coal transporters, washeries, transmission companies, etc. Perhaps, the time has come to deconstruct liability and assign it to the various actors so that vexatious litigation can be avoided.

(xi) Finally, regulators must be made accountable as much for their acts of omission as for their acts of commission. After all, the regulator is the overarching supervisor of the industry even when parts of it have been ceded to the market.

8.7 India's Regulatory Experience—Theory and Reality

In the previous section, we summarized the lessons from the case studies in this research and pointed to their policy implications. From the case studies under review, it is evident that evolution of jurisprudence in India's electricity industry, in the era of independent regulation, seems to follow a chequered path. It is not apparent that electricity regulators in India have an appreciation of their role and salience in a liberalizing economy.

In this section, we review the Indian experience of regulatory jurisprudence against the backdrop of theoretical frameworks discussed in Chap. 2. As already pointed out, most theories of regulation have emerged from Western scholarship and are generally derived in a developed economy context. While it is outside the scope of this study to develop a theoretical framework to explain regulatory behaviour in a developing country like India, we have attempted to highlight how the Indian case studies deviate from certain key conceptual frameworks that have come to define regulation.

(i) **Transaction Cost Theory: Regulation as a Means to Minimize Transaction Costs**

Of the many theories of independent regulation discussed in Chap. 2, the one that speaks to the context of this book is the transaction cost theory, attributed to Coase (1937), Williamson (1975), Goldberg (1976) and Klein (1980). This theory maintains that economic institutions arise in order to minimize transaction costs. Historically, vertical integration used to be the default industrial organization for large network utilities in most parts of the world. In India too, vertically integrated state electricity boards (SEB) generated, transmitted and distributed electricity to the final consumer. However, beginning in 1996, SEBs were unbundled into generation, transmission and distribution and corporatized. The original plan was to privatize both generation and distribution leaving transmission business under state ownership to be regulated. Coordination between all three segments would be achieved by a series of bilateral contracts between the segments. For instance, distribution companies enter into long-term power purchase contracts with generation companies, both private and public. Generators and distribution companies also enter into transmission contracts with the transmission company that owns the electric grid to transmit and deliver the electricity generated by the power plants to the bus-bar of distribution companies. From the bus-bar, the distribution company uses its own network to distribute electricity to the consumer premises. The discom has contracts with consumers for supply of electricity.

While these are not back-to-back contracts, a high degree of coordination is required between the players to supply electricity to the consumer premises. The imperative of coordination also creates huge transaction costs. Regulation emerged as a mechanism to minimize transaction costs and smoothen the coordination. In the developed world, adoption of neoliberal policies with their abiding faith in the superiority of markets meant entrusting regulation, where warranted, to commissions independent of the government. In India, and elsewhere in the developing world, the hybrid nature of the industry (private generators and mostly government utilities) necessitated that the regulator had to be independent of the government to prevent conflict of interest.

What exactly are the transaction costs that would be minimized through regulation of contracts? According to Joskow, transaction costs comprise the following: "the costs of negotiating and writing contingent contracts, costs of monitoring contractual performance, costs of enforcing contractual promises and costs associated with breaches of promises. In each case, these costs may include the cost of acquiring and processing information, legal costs, organizational costs and costs associated with inefficient (in the neoclassical sense) pricing and production behaviour" (Joskow, 1984) Each individual pair of contracting entities would find these costs substantial; an independent regulator could help reduce such costs, by ex-ante putting in place, the rules of the game, by monitoring and enforcing compliance and by penalizing infractions. According to transaction cost theorists, this is the genesis of regulation.

Joskow identifies four important determinants of transaction costs—the extent to which transactions are characterized by uncertainty and complexity, the extent to which cost-minimizing transactions require either party to make sunk investments, the extent to which there are diseconomies associated with vertical integration that must be traded off against transaction costs arising from market transactions, and the frequency of transactions (ibid).

Joskow further maintains that the greater the uncertainty and complexity, the more incomplete the contract is, i.e. it does not specify unambiguously the obligations of each party under every contingency. According to him, this gives rise to 'bad behaviour' by either party to either reduce costs or increase revenues. He further states, "transaction-specific sunk investments generate a stream of potentially appropriable quasi-rents equal to the difference between the anticipated value in the use to which the investments were committed and the next best use". This creates, he says, incentives for one party to 'hold up' the other ex-post and can lead to costly haggling. He also concludes vertical integration is the most optimal solution to minimizing transaction costs arising from forestalling 'hold-ups'. (ibid). But neoliberal policies look askance at vertical integration, especially of monopoly businesses like electric supply.

Joskow's analysis was in the context of long-term contracts that substituted vertical integration of power plants with coal supply in the USA, but it applies equally to PPAs which form the core of our analysis. Williamson and Klein maintain that where there are diseconomies of vertical integration, contractual arrangements will emerge to govern the exchange between independent agents to minimize transaction costs. The structure of these market contracts will reflect efforts to create incentives and penalties that reflect anticipated performance problems. This theory also assumes that contracts will be structured to allow least scope for legal interventions which are not only costly, but also require a degree of domain expertise on the part of the adjudicating authority. Regulatory theory tells us that an independent regulator can act as a watchdog to minimize such costs and hold-ups. It could also standardize these contracts and monitor their performance.

The relationship between the distribution utility that reaches electricity to the consumer premises and the power generator who supplies electricity is defined largely by long-term Power Purchase Agreements (PPA). Even today, power purchase from real-time markets is minimal, underscoring the importance of long-term PPAs. Section 62 of India's Electricity Act 2003 mandates that the regulator scrutinize costs of setting up a cost-plus power project and approve the generation tariff at which the discom would buy the power through a PPA, the presumption here being, the expertise of the regulator in conducting such scrutiny. Section 63 of Electricity Act deals with power projects that are awarded through the competitive bidding route and as such, prescribes regulatory approval, not scrutiny of the PPAs on the assumption that markets always throw up optimal bids.

Extant economic theory assumes that markets are competitive ex-ante (many sellers and many buyers), even prior to the call for competitive bids to build a power plant. Opportunism can emerge ex-post because certain characteristics of the supply relationship give one or the other party, monopoly power when contingencies arise. Joskow also states that the presence of durable sunk investments is the source of such monopoly power. He argues that 'dedicated assets' which are built to supply a particular buyer and thus do not have a market elsewhere would give rise to opportunism on the part of the buyer(ibid). This is also referred to as the obsolescing bargain or the hold-up problem.

The *obsolescing bargain model* describes a situation where a large corporation, usually an MNC with deep pockets and technological knowhow negotiates an agreement with a host government or its entities to deliver a product or service. However, over a period of time, the bargaining power shifts to the host government/its entity owing to large sunk investments by the corporation; such investments that cannot be shifted elsewhere or the products/services supplied by the corporation do not have alternate markets or buyers. The model surmises that the host government/entity reneges on the agreement or renegotiates it or even nationalizes the asset to extract rents. The presence of an independent regulator guards against such opportunism, according to theory.

However, in the Indian context, these theories have not panned out on the ground exactly as predicted. First, while all the projects are large and have entailed substantial sunk investments, the bargaining power has not only remained with the investor but has even got strengthened because the commodity supplied is electricity which is not only a basic necessity for a life of human dignity, but crucially, also the driver of the economy. Once contracted, the supply must go on. It is not easy to replace. This strengthens the bargaining power of the power supplier rather than that of the distribution utility.

Second, in India, thermal power generation markets are not that competitive ex-ante, considering the reliance on government for allocation for coal. There is no free market for the main input (coal) required to generate power. Coal mining and supply remains a government monopoly which is why some project developers elected to use imported coal. It is only in recent years that competitive markets are emerging in renewable power generation.

Another point to note is that in the disputes examined by this study, the investors have been companies set up by Indian businessmen, not MNCs, the only exception being the Hinduja project in Telangana. Unlike MNCs, Indian companies rely on financing from Indian banks to set up the project. Therefore, if the project developer goes into liquidation for any reason, it will have a domino effect on banks, something governments would be keen to avoid. This combined with the fact that Indian entrepreneurs usually have a deep knowledge of and familiarity with the way in which systems and institutions work in this country could be an incentive for 'bad behaviour' as defined by Joskow.

Third, transaction cost theories were conceived in the context of markets where there were no price distortions, i.e. all consumers paid a tariff equal to what it costs to serve them, whereas, in developing countries, the presence of cross-subsidies prompted by welfare considerations (alien to neoliberal philosophy) led to market distortions unforeseen by the reform designers. For instance, as the cost of power purchase goes up, consumer tariffs do not go up proportionately. There is a limit to cross-subsidizing potential beyond which tariffs cannot be raised, giving rise to what is called 'regulatory assets' (See box on Regulatory Assets) In the event, the regulator is powerless to reduce transaction costs.

Fourth, Joskow also conjectures that the contracts would be drafted in a manner that minimizes the scope for legal intervention. In view of the complexities listed above, it would be very difficult to write complete contracts that obviate the need for legal intervention. Invariably, in the Indian case, every available adjudicatory forum is enlisted in the service of adjudication of PPA-related disputes, the enormous delays and legal expenses notwithstanding.

Fifth, the PPA-related disputes examined in this work pertain to very large projects supplying substantial quantities of power, sometimes as high as a third or half of the entire power supply of an Indian state which makes them 'too big to fail'. This sets them apart in an altogether different category, investing them with additional bargaining power. Joskow's analysis would have us believe that sunk investments of this magnitude invest the host government with greater bargaining power which is quite the opposite of what happened on the ground in India.

By virtue of being 'indispensable' and also 'too big to fail' power generators have been able to make an effective case for renegotiation and win compensatory tariffs over and above what was agreed upon in the original PPA. Regulatory approval of PPAs does not seem to have prevented opportunism and the transaction costs it imposed in the cases studied perhaps because regulatory 'due diligence' was missing. Ironically, by virtue of the adjudicatory power exercised by Indian regulators, the same regulators have been petitioned time and again to revise the terms of the PPA either through amendment or through supplementary/compensatory tariffs.

While the Supreme Court has clearly stated that supplementary tariffs do not fall under Section 63 since they are over and above competitively bid tariffs, this judgement has come only in 2017 (SCC 5399–5400 of 2016). This leads us to conclude that regulatory scrutiny of PPAs that come through Section 63 route (competitive bidding) is absent and hence does little to minimize transaction costs. The assumption that market knows best and will deliver the most optimal results has done away with regulatory scrutiny although that assumption itself seems rather shaky in the light of our case study.

Our analysis points to the possibility of potential developers bidding unrealistically low tariffs, not factoring in all the risks involved in a long-term contract in order to win the contract, perhaps confident of renegotiating it at a future stage. As domain experts, electricity regulators are expected to have the competence to decide whether the winning bid is reasonable enough to ensure sustained performance over the life of the contract. And yet, the same regulators are called upon to decide on tariff revision claims.

As mentioned earlier, it is not the purpose of this research to pronounce on the correctness or otherwise of regulatory judgements on tariff revision petitions. But it is ironical that what regulators are preempted from scrutinizing ex-ante, they are called upon to review ex-post. And in the three cases reviewed in Chap. 3, the regulators have gone on to accede to the demand of the power developers for supplementary tariffs.

Ex-ante due diligence and precaution could have avoided ex-post litigation and the attendant costs, both financial and otherwise. This casts a shadow on appreciation of regulators' role in a liberalized economy.

The other point to note is that there is no consonance among the various jurisdictional authorities on the interpretation of the terms of the PPAs. In the Mundra dispute, all three—CERC, APTEL and the Supreme Court—diverged from each other in their interpretation of the terms of the PPA. CERC held that the PPA was inviolable and found no merit in the two clauses under which compensatory tariff was sought, namely 'Change in Law' and 'Force Majeure' and yet found a way to award compensatory tariffs by exercising special powers conferred by the law. APTEL on the other hand, rejected the regulator's use of special powers and held that 'Change in Law' was applicable to the case and awarded compensatory tariffs. The Supreme Court, on the other hand, rejected outright the prayers of the petitioners and held that the sanctity of contracts must be upheld. But eventually, the government intervened, setting up an 'expert' committee whose recommendations were used to persuade the Supreme Court to revise its order. The three Mundra cases lead us to conclude that eventually, pragmatism triumphed over principles.

On the larger issue of regulation reducing transaction costs, this study throws up the following conclusions:

(i) Long-term proximate relationship between the contracting parties does not seem to facilitate bilateral settlements as envisaged by transaction cost theorists. Despite the elaborate contracts drafted by sector professionals, disputes have been escalated for judicial resolution not only because the gains/losses involved are substantial, but also because the commodity supplied is electricity, indispensable to modern economic life and hence contract termination even with liquidated damages is not an option. Litigation became ineluctable in such circumstances.

(ii) The Mundra case study illustrates that where the prospects for renegotiation are bright, anticipation of possible renegotiation plays a major role in actors' initial decisions, pushing up transaction costs.

(iii) The notion that regulators are experts in their subject domain and therefore dispute resolution at the regulatory level must reach closure at that level does not seem to have validity. Dispute resolution seldom stops at the level of regulators but is invariably escalated to the level of APTEL and eventually to the highest court of the land. There are also instances where the 'catch-all' writ jurisdiction of the high courts has been invoked to settle commercial disputes. This calls into question, the acceptance of independent regulators as legitimate institutions.

(iv) The dissonance among the three adjudicatory forums on the exact meaning and implications of the various provisions of the PPA points to a lack of national consensus on sanctity of contracts in a liberalized economy.

(v) From this research, there emerge no clear-cut answers to the question of whether regulators have an appreciation of their own role in a liberalized economy. Regulatory decisions have not followed a consistent pattern both spatially and temporally. None of the speaking orders or judgements of the regulators or

APTEL reveal the actual principles informing their decisions, beyond a discussion on the meaning of the various provisions of the law and the PPA. There are instances where the regulators have taken an inflexible approach whereas the Supreme Court has taken a pragmatic approach to problem-solving. Domain expertise of regulators cannot be taken for granted.

In sum, going by the evidence available from the case studies, it would not be unreasonable to conclude that regulatory jurisprudence on the issue of sanctity of contracts as well as interpretation of contracts has not charted a consistent and clear-cut evolutionary path. Nor is the Indian experience consistent with the transaction cost approach to regulatory theory. In a liberalized economy where long-term contracting has replaced vertical integration as is the case in India's electric utilities, transaction costs do not appear to have been minimized with the advent of independent regulation. In fact, such costs have increased substantially, thanks to the tortuous and vexatious litigation it has given rise to, the Indian experience does not validate the 'transaction cost' theory of industrial organization. Whether this owes to the regulatory structure or remit or the capacity or competence of the regulators themselves or the distribution of rents from regulatory decisions, is a subject matter of another study.

8.8 Public Interest Theory of Regulation

For over a century, regulation was justified on grounds of welfare maximization. The theoretical framework for regulation assumes that natural monopolies like network utilities have a tendency to limit output and raise prices above marginal cost, leading to deadweight losses. By deadweight losses, economists refer to. suboptimal allocation of resources. It could mean pricing the product higher than what market would have done and/or not extending the connection or supply to all potential consumers. Hence, regulatory intervention is called for.

The regulator might intervene in the market also to protect consumers' interests arising out of information asymmetry between producers and consumers (Breyer 1984). An empowered regulator can demand and access relevant information from the utility for the benefit of the consumers.

Welfare maximisation and protection of public interest would warrant that the regulator scrutinise all power purchase costs. Yet India's electricity regulators play little role in the competitive bidding process. Bids to set up power projects are called for by discoms who also select the winning bid and then submit it to the regulator for approval. Whether the regulator can call into question the winning bid on grounds of tariffs being unrealistic or any other reasons is not clear, but no regulator has so far even attempted to question such tariffs. Electricity regulators have acted like a postbox, merely approving the tariffs of competitively chosen projects presented to them, rather than scrutinizing them for feasibility of sustained performance. Placing too much reliance on the market to come up with optimal prices in the hope that the contract would be performed throughout its lifetime is not in the consumer interest. The regulator should have a role to play in ensuring sustainability of supplies.

The other point to note is that although the preamble to Electricity Act 2003, the framework act which also addresses regulatory remit, enjoins upon the regulator to factor in the interest of consumers, the regulatory governance structure is not designed for consumer participation. While regulators are required to conduct public hearings where all stakeholders including consumers can participate, the capacity of consumers to actively participate in the regulatory hearings is limited. Regulators themselves, save isolated exceptions, have done little to create and groom consumer capacity to participate in the proceedings. While utilities come armed with expert legal and professional backing of consultants, consumers who are tunorganized have neither the resources nor the capacity to take advantage of regulatory hearings. Only a handful consumer organizations like Prayas and Energy Watchdog have been active in regulatory hearings, while most of the states seldom have capable consumer representation. Yet, it is the consumers who end up paying the tariffs approved by the regulators. It is fortuitous that Prayas and Energy Watchdog intervened in the Mundra cases to question the demand for supplementary tariffs. This underscores the need to integrate consumers into the regulatory paradigm.

8.9 Competition for the Market

The much-acclaimed competition for the market mooted by Harold Demsetz in his seminal paper titled 'Why regulate utilities' has also turned out to be problematic in the Indian context. Demsetz (1976) is at pains to explain that if there must be a monopoly franchise to supply a utility, it can be awarded on the basis of competitive bidding which he refers to as 'competition for the market'. According to him, scale economies do not preclude the presence of many bidders, but he makes two assumptions one of which is relevant to our study, namely "The inputs required to enter production must be available to many potential bidders at prices determined in open markets" (ibid).

As Demsetz rightly pointed out, the availability of inputs from the market at competitive prices is a precondition for the successful bidders to deliver on their contractual promise. But that turned out to be a major drawback in the Indian context. Dependent on state-owned monopoly market for coal, the bidders are at the mercy of Coal India for allocation of the fuel which is the mainstay of power plants. India's coal ministry frequently goes back on its commitment to supply coal long-term, curtailing the allocation quantities and creating uncertainty for bidders. This led to some bidders electing to use coal from own mines abroad which, as we have seen, gave rise to a different set of problems.

To take full advantage of the Demsetz model, contracts must be carefully designed. When PNGRB, India's petroleum regulator called for bids to build city gas distribution networks, some bidders quoted zero tariff although, on winning the bid, they would have to build the pipelines and city gate stations, even, install metres in customer premises and supply them. This is because, gas price is not regulated, only transportation tariff is, it enables the bidders to transfer the cost of building pipelines to the product, namely gas price. Since the winner of the bid would have exclusive rights to supply gas in the franchise for a few years, such transfer pricing is not only possible, but is also entirely non-transparent. Being a monopoly franchisee, there would be no benchmarks to compare the prices charged to consumers in order to isolate pipeline tariff from cost of gas. This defeats the very purpose of regulation. The bid designed by 'expert regulator' (PNGRB) did not foresee the possibility of transfer pricing and allowed the bidders to game the market. Eventually, the regulator had to cancel the bids. The takeaway from this case study is that regulation is not always governed by experts as the theoretical literature will have us believe.

Kydland and Prescott emphasized the importance of a framework that would produce time-consistent contracts. Other economists too favour a rule-based framework to discretion. It is the absence of such a framework in the Indian context, interpretation of contracts becomes a playground for various actors. Selective interpretation of provisions of the law or sections of the contract is used to buttress foregone conclusions. Credibility of regulators becomes a casualty in the process.

8.9.1 Competition in the Market

Economic theory informs us that even if there were no licensing or regulation by government or any authority and markets were allowed to evolve organically and function unsupervised, network industries like electricity distribution, piped water supply, etc. will evolve naturally into monopolies in their respective service areas. If this does not happen, it will be an inefficient and unstable duopoly market crying for intervention. This is because network utilities require substantial upfront investments to build the extensive infrastructure, and there are scale economies in these businesses that preclude more than one supplier. Hence, the term 'natural' monopolies.

Even if a jurisdiction begins with many competing suppliers as was the case in the USA in the early nineteenth century, shakeouts and consolidation will eventually happen and the industry will converge towards a monopoly. In the UK, Europe and their erstwhile colonies in Asia and Africa, network utilities were set up as monopolies under state ownership or some form of state regulation. This is an acknowledgement of the monopoly character of the industry that makes duplication of physical networks wasteful. Yet, the Indian lawmakers saw fit to enforce a provision in Electricity Act 2003 that allows such duplication.

Any change in market structure—such as the migration from monopoly to contestation—must be preceded by an analysis of possible implications for all stakeholders and a framework should be in place to facilitate such changeover. The raison d'etre of independent regulation is to supervise and facilitate the transition from monopoly to markets. Yet, the Maharashtra regulator let things get out of control, leaving it to APTEL to put in place, the framework for the transition.

Introducing contested retail markets through the open access route (separation of carriage from content) that the law envisages, is also problematic in the Indian context because of the existence of perverse incentives. Economic theory informs us that competitive markets deliver efficiencies through lower prices to all consumers *if and when* all consumers are willing and able to pay the marginal cost of supply or market prices. That certainly was not the case in India where cross-subsidies in tariff are rampant. Indian discoms have a mix of consumers who cannot pay full cost to serve (agricultural and economically weak sections of domestic consumers) and therefore will have to be cross-subsidized by those who can pay (commercial and industrial). This peculiarity of electricity makes it difficult to enforce open access without jeopardizing allocative efficiency. It calls for regulatory intervention in the form of levies for wheeling, banking, standby and a surcharge for cross-subsidy, not to mention compensation for distribution losses that occur in the incumbent's network. Computing, enforcing and monitoring these levies is a regulatory challenge and crucially, distorts the market to the detriment of efficiency which is the objective of open access.

Under open access, it is inevitable that the new entrants will target the viable (cross-subsidizing) category of consumers, referred to as 'cherry picking' in industry parlance. Yet, the incumbent discom has the obligation to supply everyone on demand including those who cannot pay full cost of supply. When viable consumers exit the incumbent and elect to be supplied by a new entrant through the incumbent's network, the ability of the incumbent to service its cross-subsidized consumer base gets eroded. Adding cross-subsidy surcharge to the tariff of the new entrant erodes any price advantage the latter may have over the incumbent, thus making competition unviable.

Competitive markets are based on free entry and exit. Electric supply, on the other hand, is a basic necessity whose supply can never be disrupted except for short periods. Once connected, a consumer should continue to receive supply even if the new supplier goes out of business. Especially in the case of 'open access' consumers, the incumbent utility becomes the supplier of last resort. India's electricity discoms, like all discoms worldwide, have a universal supply obligation to supply everyone who demands electricity within their jurisdiction. This obligation, an acknowledgement of the basic necessity character of electricity, is enshrined in Section 43 of Electricity Act 2003.

Another crucial reason for the chequered performance of retail competition in India is the tension between the provisions of the law and the actions of the government. Section 65 of EA 2003 categorically states that when the regulator sets the tariff of discoms, it should cover the entire cost of supply, but, if the state government would like to subsidize any consumer category over and above what is covered by cross-subsidies, it should pay this in advance to the discom through budgetary subvention. In the initial years after the reforms were introduced, state governments did make up the shortfall through budgetary allocations, but over the years, cash-strapped state governments fell back on these payments, forcing the regulator to create an entirely new unfunded category of tariffs called 'regulatory assets'. These peculiarities of the Indian context should have been considered before introducing contestation in Mumbai's retail electricity markets.

Sequencing of contestation should have preceded introduction of competition in retail supply. In the UK and EU, wholesale markets for electric supply preceded the introduction of retail competition. Wholesale suppliers bid in day-ahead power pools from which discoms source their supply at competitive prices. This is not the case in India where discoms have been saddled with long-term Power Purchase Agreements, many with pass-through fuel costs which they must honour. With escalating fuel costs and the introduction of renewables, power purchase costs of discoms have soared in recent years to account for as high as 80 per cent of the consumer bill in most states.

Even in the USA, the Mecca of Markets, regulated monopoly remains the dominant paradigm for electricity retailing. Direct competition between electric utilities that each own overlapping networks of distribution wires is quite rare. Retail competition—whether or not accompanied by duplication of wires—remains a controversial concept.

References

Breyer, Stephen. 1984. *Regulation and its reform.* Harvard University Press.

Demsetz, Harold. 1976. Economics as a guide to anti-trust regulation. *Journal of Law and Economics* 19 (2): 371–384.

Estache, Antonio, and Liam Wren-Lewis. 2009. Toward a theory of regulation for developing countries: following Jean-Jacques Laffont's lead. *Journal of Economic Literature* 47(3): 729–70.

MIX
Papier aus verantwortungsvollen Quellen
Paper from responsible sources
FSC® C105338

If you have any concerns about our products,
you can contact us on
ProductSafety@springernature.com

In case Publisher is established outside the EU,
the EU authorized representative is:
Springer Nature Customer Service Center GmbH
Europaplatz 3, 69115 Heidelberg, Germany

Printed by Libri Plureos GmbH
in Hamburg, Germany